Dave U Sloan

Fogy Days, and Now

Or, the world has changed: the innovations of the 19th century

Dave U Sloan

Fogy Days, and Now
Or, the world has changed: the innovations of the 19th century

ISBN/EAN: 9783337344771

Printed in Europe, USA, Canada, Australia, Japan

Cover: Foto ©berggeist007 / pixelio.de

More available books at **www.hansebooks.com**

OR,

THE WORLD HAS CHANGED,

THE INNOVATIONS OF THE 19TH CENTURY.

BY DAVE U. SLOAN, ATLANTA, GEORGIA.

The world moves on, it does progress,
   Rests not, rushing on, on it goes,
Where or whitherward, it may be bound,
   Is veiled, God Himself only knows.

The rage now is, to let her roll,
   Roll on, rush on, regardless where;
Let her roll, we'll cross the stream,
   Though we know a maelstrom's near.

To-day every man's for himself,
   Hindmost left to the devil's care,
The tickling game's the winning card,
   Man must tickle to get his share.

If all progress was but for good,
   Both good and evil, run along,
Side by side do their waters flow,
   But evil seems the bigger prong.

Sometimes we gaze into God's expanse,
   Peer out into a thousand years,
Then look back at the trifling past,
   And smile at former joys and fears.

FOOTE & DAVIES,
PRINTERS AND BOOK BINDERS,
ATLANTA, GA., 1891.

PREFACE.

It has been said that truly great men do not like to talk about themselves: and why should they, when their names and deeds are in the mouths of all the people?

If the little ones of the world don't speak out for themselves, how can they be heard from?

Therefore, we pray the indulgence of our readers and trust they will make due allowance for the egotistical little word "I," so frequently used in this bungling production.

In the very outset, we confess our verdancy in the art of book-making, and in taking the risk are fully aware of our liability to be cropped by the frisky kine from the herds of the literati—especially the gory ones from the clover pastures of authordom—and if any such should stray in our rural pathway, and perchance nip from our coarser tufts, the greens are not unwholesome—won't hurt them; they are welcome. Our wild grass ranges, or commons, are not intended for them, but for the people—God bless them—the best of all.

We are unacquainted with the science of music, absolutely in the dark as to its very rudiments, yet have observed when we chassa our horse-hair across the bridge of our old fiddle, that our hearers are inclined to pat their feet.

Nor do we make any pretense to erudition, elegance of diction, metrical verse, or even to grammatical sentences; but if we can only so shake our literary tambourine as to strike a responsive chord with the hearts of the people, then our labors shall not have been unrewarded.

In this little daub of a book, we have dabbled both in verse and prose, and though the structure may appear rustic and uncouth, we have tried, nevertheless, to drive the nails square in the wood, and if it should be said there is more truth than poetry in the make-up, we shall not appeal from the verdict.

Any way, let the effort be considered good, bad or indifferent, we are alone responsible; have copied after nobody. It's all home-made truck, and if critics can discover nothing but our ears, we want it distinctly understood that we have not assumed the disguise of the lion's skin.

Our book is indited to the people, and we have tried in our simple way to illustrate some of the scenes and customs of the days of fogyism, trusting to the youth of the present day, it may afford amusing portraitures of the days of "yore," and to our old-time folks suggest pleasing reminiscences of the "Days of Auld-Lang-Syne."

DEDICATED TO THE MEMORY OF MY MOTHER,

A noble, wee-bit of a woman, with an enormous heart, made up largely of love, lamb and lion; afraid of nothing above, beneath or on the earth, but God Himself, and a cyclone; a devoted Christian mother, wife and friend, with refined and literary tastes; the very soul of rectitude, and fearless champion of right under every circumstance; the faithful mother of a round dozen of children, and died scratching for her brood.

ENTERED IN THE OFFICE OF THE LIBRARIAN OF CONGRESS, AT WASHINGTON, D. C., IN THE YEAR 1891, BY DAVE U. SLOAN.

INTRODUCTION.

Have often tho't I'd write a book,
Tho' had grave doubts how it would look.
To write a book should have knowledge,
To look nice should come from college.

But I've not been through such machine.
Their inside walls have never seen,
Therefore am short in education,
So much needed in this vocation.

I'd clutch the idea, then abolish,
Because I could not give it polish.
Still it haunts from time to time,
I'll let it slide in prose and rhyme.

I'll start in verse—see how it goes—
I'll mix it up, both rhyme and prose,
The garbage may not please the eye
Of cultured critics, nor shall I try.

Out on the world the book I'll cast,
Waft out the songs of old times past;
Songs of the good old times I have seen,
What I have heard and where I have been.

Old-time happenings set afloat,
Made up of story and annecdote.
Contrast to-day with foggy times,
Show 'em up in bungling rhymes.

From the days of yore, I take my text,
Our fathers' days with ours affixed,
Warp of Auld-Lang-Syne, woof of yore,
From which will weave a cloth of lore.

Now, in this day of innovation,
In this grand, progressive nation;
Now, when these young canny laddies,
Gathering wisdom, beat their daddies.

INTRODUCTION.

E'en they, if they scan these pages,
Might glean news of other ages.
Lets now step back some fifty years;
Excuse me, please, must dry my tears.

First saw light 'midst vines and bowers,
Balmy Florida, land of flowers;
Born the time the jessamines bloom,
Born in sound of the gulf waves boom.

Birthplace, too, of the dread cyclone,
Whose life is not unlike my own,
Nativities both in south's extreme;
A brief cavort and a sorrow's dream.

Transplanted thence to Palmetto State,
Where my parents did re-emigrate.
There got my imperfect schooling,
My only plea against critics ruling.

Grew up there, and grew a Democrat,
Died in the wool, tanned in the vat;
Reared in sight of the great Calhoun
In his zenith, his high, mid-noon.

A statesman true, with eagle eye,
A man that boodle could not buy.
Like all the State to him I'd freeze;
If he took snuff we all would sneeze.

And glorious sneezes we then snoze,
Every sneeze, tighter to him froze,
In the good old days of long ago;
Blessed days, but now do seem slow.

Thought leads back to old Pendleton,
'Twas there my reckoning first began;
First hopes all hail the sunny gleams,
Steals o'er my soul like happy dreams.

Sweet to revel in memory's strain;
'Tis sweet solace to a tired brain,
But, oh! so sad, all gone forever!
Return no more forever, ever.

TABLE OF CONTENTS.

	PAGE.
Introduction	7-8
On Seneca's Banks	9
Fairfield Valley, N. C.	12
The Deer Drive	14
Land of the Sky	17
Last Hunt with Hampton	19
Cashiers Valley, N. C.	22
South Carolina Home	33
The Halcyon Days	36
The Old Slave Regime	38
The Corn-Shucking	42
The Sunny South	45
Fifty Years Ago	47
Old Pendleton, S. C.	55
John Caldwell Calhoun	62
This Day of Progression	81
An Age of Monopoly and Greed	84
The Two Streams	88
Earth's Three Epochs	91
The Pewter Buckle Moulds	93
My First Horse Trade	98
Mountain Sprouts and Sand Lappers	103
Here's Another	108
Disappointed Love	110
Shirt-Tail Canyon, California	117
Chased by Wolves in California	120
Rabun County, Ga., Frolic	124
The Victim	131
Falling Off a Mountain	138

CONTENTS.

	PAEG'
The Anxious Enquirer	141
How I Got Rid of Prince Albert	145
The Prophetic Speech	148
The Unexpected Preach	159
A Historic Horn	163
Dried Apple Cider	167
An Olden Time Fox Chase	172
The Cracker Girl	181
Prohibition Victory in Atlanta, Ga	191
Our Old Chieftain	193
The Little Purp	195
The Messenger of Peace	196
Who is Poor	197
Hotel Poetry	199
Sewing Machine Poetry	201
The Census Taker	202
Judge Bleckley's Phantom Lady	209
The Poor Boy	213
The Old North State	217
The Junius Letters	225
The Old Stone Church	231
The Confederate Soldiers' Home	235
Conclusion	238

THE FOGY DAYS AND NOW;

—— OR, ——

THE WORLD HAS CHANGED.

ON SENECA'S BANKS.

On Seneca's banks so often fished,
 Her woods and fields all I wished;
There drove the deer, knew every stand,
And chased the fox through brake and strand.

Have hunted every dell and hill,
There slaked my thirst from every rill;
From tree top did the squirrel bring,
Shot down the partridge on the wing.

Have treed the 'possum and the coon,
"Larnt" the signs from stars and moon;
Before the lark, didn't count it trouble
To hunt the roost where turkeys gobble.

Picked the strings and drawed the bow,
To lively tunes fiddle and banjo;
My old tutor darkey, Fiddler Jack,
How these memories carry me back.

Back, back to good old days of yore,
Back to the olden days galore;
To that home in the Piedmont land,
Where mountain zephyrs softly fanned.

'Back to olden days of pleasure,
The days of luck, ease and leisure;
Days of youth, when the heart was glad,
Before sorrows came to make sad.

Free as air to go, as free to come,
Bring our friends to a father's home;
Then so happy to entertain,
And ne'er to see the like again.

The rich cut-glass and old sideboard,
A custom then could well afford;
Such hospitalities then would share,
Its absence from a home was rare.

Ever with rich juices filled,
Ever stood with the best distilled;
Full welcome, never lock or key,
A jovial dram for you and me.

And the sugar loaf too was there,
Aromatic nutmeg for toddy rare;
Or fresh from garden, the fragrant mint,
Free to all, nor thought of stint.

To drink a friendly toast was nice,
Before prohibition gave advice;
Drank good cheer to friendship true,
A drunkard then scarce ever knew.

Long table spread, many a seat,
Where the welcome guests all could eat,
And merrily, merrily passed the day,
With friend and friend the old time way.

The great Blue Ridge full in sight,
So azure blue, else clothed in white,
Could view afar their craggy heights,
And oft' have clambered o'er their flights.

How much in bliss there realized,
When all for sport have sacrificed;
In summer time there was my home,
Tramped from valley up to dome.

With dogs and gun, my chief delight,
I worshipped them and thought it right;
Downed the buck in its wildest route,
Flirted from the shoal the speckled trout.

There first read the " Lady of the Lake, "
Where poet's pen did heroes make;
On rock couch had my reveries woke,
By the wierd sound of ravens' croak.

Listened to music from following hound,
Traced echoes from the horn we wound;
And it did seem heaven there and then,
If another on earth, Oh where, and when.

Mysterious world, thus to sever,
I but dream of what's gone forever;
To me it seems but yesterday,
For time at his old tricks doth play.

FAIRFIELD VALLEY, NORTH CAROLINA.

Vale of Avoca, have never seen,
But Fairfield would beat it, I ween;
On a knoll, in that lovely vale,
Sat our cottage, gem of the dale.

By a brooklet so chrystal clear,
Sky-scraping mountains in the rear;
East there flowed the cleanest river,
As limpid as the Gaudalquiver.

This river's name was Toxi-way,
Doubtless is running there to-day;
And there sported the spangled trout,
'Twas my delight to lift them out.

Across the river, mountain chain,
Making off from the Blue Ridge main,
To the left, and also parallel,
Walls of blue rock, remember well.

As rounding high, a thousand feet,
Which do, too, the Blue Ridge meet;
And Fairfield Valley lies between,
Nor fairer vale was ever seen.

In contour oval to the eye,
And level doth its bottoms lie;
The river heads north, full in sight,
From a thousand rills with waters bright,

OR, THE WORLD HAS CHANGED.

Comes dashing down into our vale,
Making a river of silvery trail;
It's egress south, does seem shut in
As if no outlet, none had been.

Ridges seemed joined in solid wall,
But through a gorge the waters fall;
Go plunging down this narrow way,
Mad'ning in their boisterous play;

Reckless leaps to the dell below,
Plunging, foaming, and white as snow.
Just down there, once, we had a mill,
Wonder if it grinds, sawing still?

High, o'er cottage, a mountain top,
As if upon perchance might drop;
To the north, standing stark and stiff,
The mountain backbone, grand Sheep Cliff.

And mountains circling all around,
So was this lovely valley bound;
Much like some great amphitheater,
Built by God, the grand Creator.

A scene so grand, indeed so great,
Artist hand dare not imitate;
And is so fraught with Nature's gush,
The tints must come from Heaven's brush.

The sun climbs o'er the hills at ten,
Shines o'er this deep basin—and then
Hides its head at four, sinking down,
Shuts out the curtailed horizon.

Another valley, across a gap,
Cashier's, and lying like a lap;
The lap of this great mountain chain,
And lies there yet, if been no change.

These valleys lie there side by side,
Their beauty no one hath denied;
And here our Southern people came,
Some for health, some in search of game.

They came in search of game and health,
All of means and many of wealth;
Here they came to spend their summers,
Were attracted there, many comers.

These valleys in the old North State,
There yet, if not removed of late;
Very near its Southern border,
Sure we left them in that order.

From the State of Buncombe were due West,
These valleys of so much interest;
Left them there hanging near the sky,
Among the clouds ahanging high.

And here we gathered every Spring,
Our guns and dogs along would bring;
Came to enjoy each one full share,
So back behind we left all care.

THE DEER DRIVE.

Would break our fast at early morn,
Called together by signal horn;
When eager hounds with business yelp,
Seemed crying, "Masters, here's your help."

Horn answers horn to sound the meet,
We'd start the hunt before the heat;
Then off to the wilds we'd repair,
To roust the game from out his lair.

First plan the drive, start in the hounds,
Then post the standers on their grounds;
All ready, each one for his part,
The drivers in to make the start.

Anon, we hear the shrill halloo,
Down in the cove, way down below;
Watching, listening, catch a sound,
'Twas but loose tongue, a puppy hound.

But, there again, old Troop strikes trail,
Troop is true, never known to fail;
There's Haidee, too, and she's a blood,
They now give tongue all through the wood.

Aye, aye, and now have sprung the game,
The pack all in, and are aflame;
They follow close, the scent is strong,
Now the grand chorus swells along.

Lookout standers, now watch your ground,
Ha! Here they come, the crying hounds;
What bodes the weakness in my back?
The tremor doth my legs attack.

Have heard, if symptoms don't deceive,
Case of buck-ague, we do believe;
LaGrippe has got us in the back,
Has got us, got us, fur a fack.

But they have tacked, turned another way,
And our chance is lost, lost for to-day;
The game is wary, plays around,
Can trace him by the following hound.

Other stander may be in luck,
May be his day to kill the buck;
Nervously we watch, watch and wait,
Hi! They come again, coming straight.

A crash, a thug, our ears assail,
See there branching horns, cotton tail;
Quick bounding past, as like a streak,
And right now's our time to speak.

Bang, bang, our double-barrel went,
And two buck loads at him we sent,
Aha! We see he drops, drops his tail,
His agile spring begins to fail.

The pack sweep by, as like a storm,
They scent the blood and make it warm;
And, like the wind, they follow fast,
And like a cyclone now have past.

'Tis all over, the dogs at bay,
Glory enough for one short day;
The chase is ended, the stag is dead,
The hounds around are gather-ed.

Now in triumph, filled with pride,
The dogs at rest, are satisfied;
Now we sound the gathering call,
Answer winds back from one and all.

Have heard the signal and obeyed,
And up rides the jolly cavalcade;
So went the hunt from day to day,
If not the same, then another way.

Sometimes rewarded with a bruin,
Or sleek panther, a beast of ruin,
Wolf or catamount, all the same,
For we were out, our purpose game.

Turkey and pheasant we often shot,
To grace the table, fill the pot;
So all our summer days were spent,
For, like business, at it we went.

OR, THE WORLD HAS CHANGED.

LAND OF THE SKY.

That beautful land of the sky,
Grand mountains rivalling Italy;
From their high tops the grandest view,
For vast expanse we ever knew.

From great Sheep Cliff, the main Blue Ridge,
Long, narrow, like a mountain bridge;
On that high perch we've often stood,
And gazed afar 'pon field and wood.

O'er tops of hundred circling peaks,
And endless coves and cliffs in streaks,
Boundless forests, with much Spruce pine,
And all the brooks, the Laurel line.

Here and there we see bright cascades,
Snowy waters leaping to the glades;
Northward the Smoky Mountains blue,
And noted for this special hue.

Here the balmy Balsams intervene,
Wrapt in their softer velvet green,
Though together so closely linked,
Their shades of color quite distinct.

Now mark those ridges taper down,
Towards the plain, the level ground;
There we have the ocean view,
Those white spots like white caps too.

See the field and wood sink and swell,
Doth imitate old ocean well;
Watch the glories of setting sun,
Painting resplendant the horizon.

Painting, guilding with such bright sheen,
Language fails, can't describe the scene;
Now feel the need of education,
Subside, no further explanation.

In Heaven we ken, mountains fair,
Grand ranges, ever standing there,
For they display God's mighty hand,
Majestic mountains bear His brand.

Silvery streams and chrystal bright,
Rivers in which the saints delight,
Who, forever sing out their thanks,
Tramping gems that line their banks.

Plucking fruits that forever ripe,
And ne'er and ne'er a tear to wipe;
No anxious thought about to-morrow,
Where the Son of God shuts out sorrow.

LAST HUNT WITH HAMPTON.

But Summer's past and Fall has come,
  Now turn our thoughts to going home.
Here's a yarn, some may call it luck,
Col. Hampton wished a deer—a buck.

A whole buck to his home to take,
So we did the arrangement make;
Take to Columbia on his return,
He'll testify to the whole concern.

Now knowing where a fine deer lay,
On Nix Mountain, there let him stay,
'Till by appointment, when enroute,
Had laid our plans to get him out.

Ready, sent our negro driver,
If this buck he could diski-ver,
And we rode round to hold the gaps
Where the game would pass, no mishaps.

Nor did we have to daily long,
Until we heard the dogs give tongue,
In a jiffy the game was sprung,
And at his heels the dogs were strung.

The deer made direct to our stand,
Double-barrel cocked in our hand;
Our eye fixed on the coming game,
And we were nerved for deadly aim.

Crash above called our attention,
'Twas the Colonel making our direction,
Galloping down the mountain side,
As rapid as a man could well ride,

Came quartering toward the deer,
As to intercept it did appear;
The buck was stretched at full speed,
So it seemed was the Colonel's steed.

Then we saw a blaze from his gun,
And as quick as thought another one;
The buck came on like thunderbolt,
As if shot out from catapault.

And fell dead within twenty rods,
That shot was worthy of the gods;
Wagon was waiting at the road,
That buck made part of the Colonel's load.

The last hunt with him we ever took,
Just as we tell it in this book.
Col. Hampton then was young and rich,
A full made man in every stitch.

A man who no one bore ill will,
A hunter bold and one of skill;
A soldier born, though then untried,
Now known to fame, far and wide.

Of the best timber was he made,
And braver ne'er donned the plaid;
Nature's nobleman, luck or adversity,
The hero be known to posterity.

———

HAMPTON, MISS., April 16th, 1891.

MY DEAR SIR—I have been traveling about so much of late that my correspondence has fallen in arrears, and thus your kind letter directed to Washington remained unanswered.

I remember well the incident you refer to, as I do many pleasant hours spent with you in the mountains of North Carolina.

There have been many changes since those days, and many of them for the worst, but I hope that our South may yet be prosperous. With my kind regards I am,

    Very truly yours,

        WADE HAMPTON.

P. S.—Two or three years ago I shot a buck here which weighed with entrals out 265 pounds; his skin, from neck to end of tail, is seven feet long. I have here, too, a pair of horns with twenty-eight points.   H.

To D. U. Sloan, Atlanta, Ga.

CASHIER'S VALLEY, N. C.

This valley was named for a horse, that strayed from its owner, James McKinney, of South Carolina, and after months was discovered grazing in security there. McKinney was so well pleased with the locality that he afterwards settled there, and spent the balance of his life in Cashier's Valley.

Situated upon the very apex of the Blue Ridge Mountains, in North Carolina, this valley is one of the most elevated in the State, having an altitude of near 4,000 feet above tide-water, bounded on the north by Sheep Cliff, the backbone of the Blue Ridge, east by the Rock Mountain and Chimney Top, south by the Terrapin, and west by the great Whitesides. Passing through McKinney's Gap, to the north, crossing the ridge, one would descend into the fertile Tuskaseege Valley, and crossing a gap to the east, would drop suddenly into Fairfield, 300 feet lower than Cashier's, and one of the most beautiful valleys in all this range; going west, would enter the valley of Horse Cove, nestling under the shadow of the Whitesides; following the waters south, would be brought to a sudden halt by the White-water Falls, equal in volume of water, and vieing in its magnificence of scenery with the famous Tallulah Falls of Georgia. After several stupendous leaps, this beautiful clear water stream plunges into the Valley of Jocasse, S. C., where, uniting with other streams, it forms the Keowee River.

Passing through Cashier's Valley is a turnpike road, built before the war, by Colonel Wm. Sloan, to the North Carolina line, and from thence by Colonel Wm. Thomas, across the Blue Ridge, and on down the Tuckaseege. Cashier's is comparatively level, and the tourist could imagine he was in a champaign country, but for those huge domes that stand like grim sentinels encircling the valley, and upon every hand an eye of taste could select the most charming spots for residences. Years ago, when hunting game, and fishing for the speckled trout from their silvery beds, we would conjure up in our minds vision pictures of enchanting grounds and imposing edifices, consequent upon the advent of railroads; and now, since the development of the Richmond and Danville Railroad—the great Piedmont line—our youthful fancies have in part been realized, for a few miles away the thriving town of Highlands has sprung into existence, on one of the most elevated plateaus of the Blue Ridge—is a charming place, and is becoming a place of resort in the summer months.

The visitor, in ascending this mountain region, notices the wonderful change in the atmosphere, its bracing effect on the system, the feeling of freshness and delight experienced in this altitude. The effect on the appetite is remarkable; first keen, then ravenous. We can never forget our first visit to Cashier's Valley, our relish for old Aunt Sally McKinney's "yaller-legged" chickens, fried so brown, and floating in the golden melted butter, snow-white smothered cabbage, mealy Irish potatoes, cracking wide open as they were lifted from the kettle, buckwheat cakes and mountain honey, nor shall we try to erase from our memory old Mr. Mac's mountain dew that sat out on the water-shelf before and after and between meals.

To describe this romantic region would require the pen of

genius, and books, and after all would have to be seen in person to be appreciated. This section abounds in game and affords many delights to the sportsman. Cashier's and Fairfield Valleys were for many years the resort of some of the best citizens of South Carolina. It was here the noble Hampton loved to come out of the Summer's heat to chase the deer and catch the mountain trout; and, long before the war, the Hamptons, Prestons, Calhouns, Haskells, Chevises, McCords, Taylors, Palmers, Stevens, Whitners and Sloans spent their summers there.

> " Tis distance lends enchantment to the view,
> And clothes the mountains in their azure hue."

> The poets picture here is but partly true,
> True, its distance makes the color blue,
> Why not as pretty, if the color's green,
> As arrayed in its lovely Summer sheen,

> On near approach, the color changes hue,
> Refreshing green takes the place of blue;
> But the distance part we would refute,
> In our survey, would rather be more minute;
> To us the enchantment is in being there,
> At least we'd choose it for our share.

Viewing these mountains from the Piedmont Road, from the many glimpses to be had as it skirts along its base for one hundred miles, is a sight that must ever attract attention, but for real enjoyment the admirer should go in their midst, ramble the valleys and climb the heights, trace the dashing streams, behold with his own eyes the cloud-capped peaks, view the broad expanse of country to the south, stretching as far as the eye can reach, 'till earth and sky seem to kiss each other; see hill and dale, like ocean's undulating waves, landscape dotted with farm and villa, resembling much the white caps of the sea, and all around see monster peaks, peeping over the heads of the others, as if to catch a better view of the great world below, but, ever quiet and courteous, they do not, human-like, trample upon each others toes and cry out "down in front" to those who happen to impede their view.

These mountains possess a labrynth of magnificent scenery, and the lover of the romantic and picturesque can here revel in such delights. If one would witness Nature's grandeur in an adjective degree, let him stand on one of those towering pinnacles and watch the storm king as he sways the earth below, standing serene beneath the brightest rays of the sun, he may see the white clouds roll far beneath his feet, listen to the rushing winds, without a ruffle of his hair, hear the thunders reverberating boom, see the flashing lightnings as they cleave their zigzag course, and whilst torrents deluge the earth below, he stands dry shod, or let him peer into the abyss below, from the brink of some precipice down into the giddy depths, where houses look like toys and men and beasts as flies that creep on the wall.

If the visitor be an artist, there is a world of material for his crayon; if a poet, in the midst of these mountains is the home of the Muses; if an orator, here let him choose his rostrum and spout, for inspiration must seize him here; if a statesman, here let him climb some mountain dome, adjust his glasses, and he will, perhaps, see further than he could from the halls of "our fathers;" if a lawyer, let him come here and rest from strife and enjoy that peace he would not allow his neighbors; if a doctor, he may come here and find the roots, dig and pack them home, for no physician is needed here; if an invalid, let him hie to the Highlands, for every spring here is a Ponce DeLeon and filled with life ; if a lover, let him bring hither his charmer, seek some sequested dell, and whilst reclining on some mossy couch, with finger tips rippling in the gurgling rill, and though the heart speaks most when the lips move not, the tale would soon be told; if a Nimrod, here is the hunter's dreamland, the paradisical hunting grounds; if he has a soul for delightful chords, here he may bend his ear and catch the cadences as they fall from the musical echoes of the many-tongued pack, as it swells in song, publishing where the fleeting stag has sped; if he be an emigrant, here he can come and buy a home, rich and cheap, here is the place to feather a nest, sing "Home, sweet home," and compose new lullabys for generations yet unborn, who shall surely flourish under these Italian skies.

In this delightful region your scribbler was delighted to spend many of the summer days of his early manhood; here first read the beautiful poems of Walter Scott; here first brought down the antlered buck, and flirted from their crystal beds the golden spangled trout; here quaffed from the glassy brooks delicious draughts that ice would spoil, and breathed an atmosphere so pure and bracing that physical exercise seemed attended with no fatigue, and often since, when burning with miasmatic fevers in Southern Georgia, in fitful dreams have wandered here again, almost tasting the precious boon, when some tantalizing fate would snatch it away.

The inhabitants of these mountains are rude and illiterate, but warm-hearted and generous to their friends; they have no idea of caste, and are profoundly impressed with the idea that one man is as good as another, if not a little the best, provided he is honest. They entertain supreme contempt for the lower country and city folks, who were too ignorant to course a bee tree or follow a wolf trail, who asked silly questions and could not tell a deer track from that of a hog or a sheep, who knew nothing of signs and shot-scatter guns, who wore starched shirts and combed their hair.

Even the women regarded the men from the lower country as effeminate, and on one occasion a buxom mountain lassie bantered a South Carolina hunting party for a foot race, offering to take the biggest man they had on her back and beat their best runner. On another occasion this same heroine was seen coming from a mountain george, with her rifle, sleeves rolled up and bloody arms, and, upon being questioned, indifferently replied that she had "jist kilt a bar beyant the Terrapin."

Many years ago, Mr. John C. Calhoun, Col. Gadsden and Col. Wm. Sloan, were surveying for a railway pass through these mountains, and whilst the subject of railroads was under discussion in the presence of the mountain family, a young daughter, the pride of the household, put in: " Uncle Jim says ef he war to see one of them relerodes acomin', he'd leave the world and take a saplin'; Dad says, ef he seed the dern thing he'd drap rite down on the yeath." But now these people can mount any of their neighboring heights and watch the wreathing smoke as it curls up from the iron horse, speeding along the Piedmont hills. Long ago a marketing party from this section, with their wagons, made the great trip to Augusta, Ga. They belonged to the Hardshell persuasion and everything moved nicely 'till they got to Augusta, when one of the brethren got too much of the o'erjoyful, fixed up in ice and sweetnin', lost his gauge, and was picked up by some of the party in a gutter. After the return home he was dealt with by the church. The good brother made an honest confession, and humbly besought forgiveness for the not uncommon offence; he plead in palliation for the slip, that the ice and sweetnin' in the licker had fooled him, when one of the breth-ren exclaimed, "Stop right thar Brother Wilson; did you say they put ice in your licker?" Turning to the other breth-ren asked, "Wernt it in July we was thar?" Brother Wilson said he knowed it wer in July, but they surely put the ice in the licker; the brethren looked grave, and, after mature deliberation, decided to expel Brother Wilson from the church, not for getting drunk, but for telling a lie and sticking to it, for all the brethren knew it was cooler on the Blue Ridge than it was in Augusta, and the oldest man in the settlement had never seen ice in July.

On another occasion a mountain preacher was explaining to his audience that morals alone could not take a man to heaven; as he proceeded with his argument he became more and more convinced of the impossibility of the thing; suddenly paused for a moment, then raising high his brawny arm brought it down with sledge-hammer force on the candle board, exclaiming, "No, no, my breth-ren, the thing can't be did; you might as well tell me that a hawk could knock feathers out of a terrapin."

Another, a minister, was illustrating the meekness of the lamb, and made the following graphic picture: "Thar was once a goat and a sheep acrossing of a log and it so happened that they met right in the middle; the water was swift and the log was high; they couldn't pass one another, the log was too narrow, they couldn't turn around for fear of falling off, nur they couldn't go through one another, and to hit that bilin' water down below were surely death to a goat or a sheep. Now my breth-ren, right thar was a dilema wan't thar? Didn't it look like thar was a dilema thar? Well, thar wernt. No, my brethren, thar wernt narry need to be a dilema thar. I'll tell what they done, why the sheep squatted down and the goat jumped over, and right thar was the meekness of the lamb; and oh, my dear breth-ren and sisters, thar's a way out of every trouble; its to squat, why squat, jist to squat, in the name of my lowly Master squat, git down, git down; oh, my dear breth-ren, be ready always to squat, when the dilemas of life come—be ready to squat.

PREACHER REID.

Brother Reid was an uneducated mountain preacher, could read the scriptures with difficulty, but he thoroughly understood the plan of salvation, and could illustrate it with great force to his people, was as solid and orthodox in his principles as the rocks that surrounded him. He was to preach in Horse Cove, where Judge Whitner, from Anderson, S. C., was spending the summer.

He was informed by some of the brethren that the great Judge Whitner had come out to hear him preach, that he must do his very best on that occasion.

When Brother Reid rose to line out his hymn it could be plainly seen that something was weighing heavily upon his mind. He started to read, then hesitated, then stopping short and looking around most solemnly at the people, said: "I have been told by some of the breth-ren to do my best to-day, for the great Judge Whitner has come here to hear me preach.

I have never seen the Judge, but have hearn of him, and I suppose he is a powerful high larnt man, and if he has come to hear me preach it don't make no difference to me. Breth-ren when I preach the gospel I preach Jesus Christ and him crucified, and I don't care if Judge Whitner or Judge Thunder is here, and if Judge Whitner is in this house and is a Christian man I want him to get right down on his knees and pray with us." The Judge complied and he became a great admirer of Brother Reid and his lifetime friend.

COL. WILLIAM SLOAN.

SOUTH CAROLINA HOME.

But Fall is here, we homeward bound,
   Like the swallows that homeward fly,
Back to our homes, the Piedmont lands,
   Farewell, Blue Ridge, 'till bye and bye.

Long Seneca's gently flowing stream,
   Lies a valley most passing fair,
Valley rich with alluvial lands,
   That have long been tilled with care.

In the good old South Carolina,
   Where Twelve-mile Creek and Keowee,
Mixing waters make the Seneca,
   Below this fork dwelt our family.

The old home was named Tranquilla,
   Across the river was Fort Hill,
Famous home of John C. Calhoun;
   Both these old homes are standing still.

Across the Keowee, fronting too,
   Once a palatial mansion stood;
John E. Calhoun did there abide,
   And rich in slaves, field and wood.

West, and below, Uncle Tommy Sloan,
   Lewis, Cherry, Earle and Maxwell's,
Generals Pickens and Anderson,
   North, above, Lawrence and Liddells.

Still above, the Ramseys and Reids,
  Broad acres each and all possessed,
Their rich bottoms lined the river,
  With abundance all were blessed.

To say they lived would fail to tell,
  Larders full, graneries bulging o'er,
Lived as nighbors ought to live,
  Their latch-strings all hung out the door.

Their big white mansions crowned the hills,
  Had comforts that ne'er can be told,
Though tedious in this little sketch,
  We try to tell of the times of old.

Each house, at times, was made hotel,
  And oft with friendly guests was filled,
Who came in squads and families,
  And who oft for days were billed.

Friends would come from many a mile,
  In old-time coach and baggage carts,
Children, servants, sometimes their dogs,
  And they would come with jolly hearts.

Then fowls and swine, and fatted calf,
  So freely slain on their advent,
And warm welcome greeted everyone,
  Such welcome as was surely meant.

Would feast and frolic, and entertain,
  So the old-time days flew by apace,
Played old-time games, as blindman's buff,
  The boys would jump and run foot race.

Their daily sports to hunt and fish,
  The nights made merry with the dance,
Neighbors called to swell the throng,
  Help out the fun and hold parlance.

No morn nor eve allowed to lag,
  Each day had its own sensation,
Fun and frolic was in the wind,
  Those old-time days of recreation.

They had good times throughout the year,
  But Christmas capped the climax,
Every door thrown wide open then,
  Hung with mistletoe and smilax.

Santa Clause ne'er failed to come,
  Nor the little ones, unbefriended,
Hanging stocking ne'er failed to fill,
  For he ever came full-handed.

Time was not counted money then,
  Lucre needed, sires freely gave,
Would run accounts with all the stores,
  And never sought a bill to stave.

Each family had yearly account,
  Everybody would do to trust,
Tho' prodigal the total amount,
  The money due, down came the dust.

Ah! Those were rosy, daisy days,
  But now have gone, gone glimmering,
Love to think of those happy days,
  Whilst our life away is simmering.

THE FOGY DAYS AND NOW;

THE HALCYON DAYS.

Since then have seen a bit of the world,
    Had trials, seeming of perdition,
In all our wanderings never known
    A people happier in their condition.

Intelligence, freed from fashion's chains,
    Wealth, divorced from aristocracy,
When heart was free to show its hand,
    'Twas time of the true democracy.

Such was confidence, that money loaned,
    'Twas rarely asked to give a note,
Promise was better guarantee
    Than legal paper daftly wrote.

Don't claim there were no rascals then,
    That would be to cheat the devil,
Admitted fact since Adam sinned,
    There has been more or less of evil.

There was a social line and plummet,
    Character weighed more than gold,
The man who did a dirty trick
    The good people would not uphold.

They despised that narrow leanness,
    The Yankee grip on a quarter,
Happier to bestow than receive,
    And their favors flowed like water.

Did not consider swindle smart,
    Did not seek to rob a neighbor,
Would not lose sleep to undermine,
    But for the right they would labor.

Bible folks, too, that is mostly,
  Humanity never was perfected,
One cheek smote, the other turned,
  Then it was that they reflected.

Winked at the old code duello,
  Where there could be no compromise,
Where blood or life must wipe out
  Stains 'twould not do to temporise.

And to-day 'tis an honest querry,
  If 'tis not best mode of settlement,
Where honor is so deep involved,
  To make a test of mettlement.

Now looked upon as barbarous,
  This great age the code discarded,
Old times belonged to refinement,
  To best classes was awarded.

'Twas handed down from chivalry,
  Days of Fitz James and Roderick Dhue,
Of bonnie Scotland's literature,
  Doubt if we, better type, don't you?

Our old time folks had fogy ways,
  A cowardheartily did despise,
Thief, even too low for contempt,
  And a liar would soon assize.

In truth, they stood on higher plain,
  Their social order less impure,
Less of deceit than this fast day,
  And in virtue far more secure.

The war has brought its bad results,
  And our old customs it has changed,
Are learning fast the Yankee plans,
  From the right we have been estranged.

THE OLD SLAVE REGIME.

Our paternal grounds spread out wide,
View them o'er took many a stride,
Were owners too of many slaves,
Who worked our crops, dug our graves.

Cooked our food, brushed our suits,
Hitched our teams, blacked our boots,
Hauled the wood and made the fires,
And did these things for our sires.

Kept a dozen about the house,
Some in livery, some in blouse;
Fed and clothed, thrashed them well,
If they got too mean, then would sell.

Master was good, if slave was true,
And such were nicely treated too;
Did as well for them as we could,
Doubtless as well as Yankees would.

Gave them tobacco, sometimes dram,
These were the things that tickled Sam-(bo),
And made his white teeth to shine,
When well pleased this was the sign.

Their little cabins were all in line,
Like little town, but not so fine;
'Twas what we called the quarter,
Near some spring, or running water.

Rations per week, a peck of meal,
Measured correct in high state seal,
Besides a pint of molasses,
They provide all other sasses.

Three pounds bacon, woe or weal,
Thribble as much, if beef or veal
A good patch, too, each family had,
With right to work it good or bad.

Saturday's gave them half the day,
To work their patches or to play,
And at night they would pat and jig,
Monday morning its plow and dig.

Some had chickens, a pig and a cow,
How many nigs doing better now?
Some were lazy, some had thrift,
Some would work, some had to lift.

A Sunday rule, they must come out,
   In their best suits were to be seen,
Their kinky heads be carded up,
   Sunday must show up neat and clean.

Sometimes he'd rob master's roost,
Sometimes master made him boost,
And then again he'd run away,
But then again 'twas master's day.

And Sambo said, more rain more ress;
What, sir? "I sez more rain more grass."
Ah, said master, pretty well done,
You rascally African son.

One peculiarity is his scent,
Whene'er he moves he gives it vent;
After all has been a useful race,
After all, a good thing in its place.

Once Sambo to the Lord did pray,
Master Lord, let me die dis day;
Wicked boy ensconsced o'er head,
Cried out, Sambo, and to him said:

Come, Sambo, your prayer is heard,
Come home, the Lord has got your word.
Who dat, cried Sambo, who sed so,
Dat you marse angel, call Sambo?

Done come for him, is dat you say?
Why he done dead de lass tree day;
Marse angel tell em him done gone,
Yessa, he done dead, dis same one.

A truth that claims recognition,
Their one great trait, superstition;
For take them one or by the hosts,
As a race, all believe in ghosts.

Southern slavery may have been sin,
   But the Bible does not show it;
There is abuse in everything,
   If forbid, would like to know it.

By Yankee means he now is free,
   We believe the Lord hath done it;
His days of bondage had run out,
   By the powers above he won it.

Flames first kindled by Madame Stowe,
  Crazy John Brown then made it roar,
On to the bloody shirt to-day,
  Flaunted by grannies Ben and Hoar.

May be the Lord to Christianize,
  Allowed Yanks to do the stealing;
Sold the niggers, then felt the sin,
  Got the pay, then did the squealing.

And so the nig was the winner,
  But our Yank he got the credit;
Poor Dixie was a cat's paw made,
  And to the Rebel falls the debit.

But slaves were happy in the main,
  Of course, exceptions in all cases,
No heavenly state here below,
  'Tis not in reach of earthly races.

Let's have old Sambo take the stand,
  Let old time nigger tell the truth,
Which times were best, freedom times,
  Or the old slave times of your youth.
Were you happier then or now?
  Give us truth, weigh upon the scales,
As slaves were you not free from cares,
  Your only fears the lash and sales?

Didn't love poor bucra overseer,
  Your terror was the patter-roll;
To leave your home must have a pass,
  Or risk your heels to save your poll.

Have you forgot your little thefts,
  Of all the chickens you have stole,
Of the tater patches you have robbed,
  Couldn't count them for your soul.

You had your faults and had good traits,
    Were faithful in our distress,
Were true to us in time of war,
    Left in charge of our business.

A happier people ne'er was known,
    Than the old-time Sunny South,
Including slavery in its bonds,
    Subject of so much Yankee mouth

THE CORN SHUCKING.

But best of all he loved to sing,
    And in song indeed was gifted,
The field and wood he made to ring,
    Belly full, into song he drifted.

His gala time—the corn shucking—
    No cards needed to bring him in;
As sun went down could hear him shout,
    For he was coming to the binn.

When full gathered, a motley crew,
    They would come from many a mile,
Without regard to sex or size,
    Would gather around a corn pile.

First choose their leaders for the fray,
    And then the leaders pick their sides,
The pile of corn is struck in half,
    Over which each captain now presides.

Word given then a rip of shucks,
    The ears go flying o'er the pile,
Shucks are pushed back to the rear,
    The captain cheering all the while.

Each leader walks on top the pile,
    Midst the showering ears of corn,
They walk and shout and lead the song,
    And far away their songs are borne.

The war waxes to fury fast,
   'Tis strife, who shall the victory win,
The pile grows less at every turn,
   No fiercer fight, tho' thousands slain.

All through the clash the jug goes round,
   From mouth to mouth the goody went,
As fast, faster the corn would fly,
   'Till the unshucked corn was spent.

'Tis then the victors heave a shout,
   A shout that rends the very skies,
Now the devil seems turned loose,
   And its now the master flies.

For the boss is seized if found,
   Is hoisted o'er the darkey heads,
With shout and song they bear him round,
   To where the supper table's spread.

In home yard, on rude table laid,
   Is fowl and shoat, and lusty pies,
'Possum and 'tater, many a dish,
   Canopy o'er head, God's blue sky.

And next the fiddler thumps his strings,
   A dusky crowd round pine torch light,
And dance with all their might and main,
   Regardless of the fleeting night.

There never was a happier race,
   If they could have been left alone,
'Twas hatred that stirred up the fuss,
   The Yanks were jealous of our bone.

To solve their future, the problem,
   One intricate to unravel,
Shall they stay? Must they go, or no?
   We think they will have to travel.

'Tis the great question of the day,
   He has already cut a figger,
He'll never ride the upper rail;
   But just now we'll drop the nigger.

THE SUNNY SOUTH.

Great mistake think our fathers made,
　　Raised their sons without work or trade,
Raised as gentlemen, were not prepared,
If had been trained, had better fared.

In those evil times that were in store,
In the troubles that tried them sore,
They felt the keener that distress,
Consequent upon their idleness.

Were taught in honor—that was well;
But that alone doth not propel;
They learned aptly how to spend,
This was their trouble in the end.

Wrong idea of the old time South,
　　Thus a noble generation was lost;
Ill prepared to meet and grapple,
　　They have learned at heavy cost.

The world was not made in a day,
　　Takes longer to make a nation,
And blood that tells take time to breed,
　　Must have culture and recreation.

The works of time may be impaired,
　　Noblest monument may be marred,
The grand old oak may be despoiled,
　　And its rootlets all be scarred.

So the grand old South long had stood,
　　Its great branches were spreading wide,
A brother's axe hath cut it down,
　　And e'en prostrate they still deride.

Nor spared in hatred, yet pursue,
　　E'en in defeat they still would vex,
And seek to hoist an accursed race
　　To place their feet upon our necks.

They shirked the slave off on us,
　　Because they could not make him pay,
Then again they were dissatified,
　　Have robbed us, stolen them away.

In their zeal, 'twas "snake in the grass;"
　　We do not speak in hate or spleen,
We do not wish the Yankees harm,
　　We do not think they all are mean.

They made their money out of us,
　　We hewed the wood, drawed the water;
Our good friends when served their ends,
　　Gave big end in every barter.

Now we are glad the negro's free,
　　Tho' 'twas hard at first to swallow,
Has broken up old fogy plans,
　　In which we were want to wallow.

Now we grow, even more than they,
　　And in progression shall compete;
We'll make our cotton into cloth,
　　Thus their own plans will defeat.

The South will run her factories,
　　Run them for all the money's worth;
Tariff paid them will keep at home,
　　We will have the new South henceforth.

'Tis a long lane that never turns,
　　Hair from dog is good for the bite,
Just keep still, and things will turn round,
　　Darkest hour comes before daylight.

Blood will tell, tho' it seemeth dead,
　　Will rejuvenate, will flow again,
Scions will spring and flourish here,
　　Tho' the paternal stalk be slain.

New scions shall take the firmer root,
　　True scions from a noble race,
Who shall be rulers of this land,
　　No darker blood can e'er displace.

For 'tis written in their very hearts,
　　Written there in blood's red ink,
'Twill never be recorded here,
　　We are ruled by a race that st-kink.

Let Southern States as sisters be,
　　True sisters walking hand in hand,
Their native worth is sure to win,
　　There's none like them in all the land.

Like lilies bent by stormy blasts,
　　And as the eagle stoops to rise,
Fair Dixie thou hast but to wait,
　　For thou shalt soar as the eagle flies.

FIFTY YEARS AGO.

Fifty years ago, age of content,
　　Before fashion's laws were defied,
And our worship was so simple then,
　　When our wants were not so amplified.

Walked in the paths our fathers trod,
  In suppliance bent the humble knee,
Took our dinners to the meeting house,
  And was so glad each other to see.

In those frugal days our wants were few,
  The only fashion was to be neat,
Didn't care much for outside show,
  But sure have something good to eat.

Better days than now, it seems to us,
  Although didn't know near so much,
And some things are glad we didn't know;
  Indeed, would have been ashamed to touch.

Before the day of the patent pill,
  Days of the lancet, the calomel,
Doctors didn't try to size your pile,
  But worked harder to get you well.

And justice was better meeted out,
  Tho' the lawyers were not so plenty,
And neighbors were less at logger-heads,
  Not so many suits, not one to twenty.

Our preachers then were humbler, too,
  Like Paul, labored for their living,
Preached because they loved the Lord,
  Wan't so rantankerous 'bout giving.

At church they sang most sacred songs,
  To the old time fogy meter,
And these old songs didn't make you feel
  As in a circus or theater.

Sang hymns to the old familiar tunes,
  Were sacred in all their bearings,
Could not mistake it for caterwaul,
  For naughty felines on a tearing.

'Twas antideluvian as to choirs,
    Had no big bellowing church organs,
And the congregation sang God's praise,
    Did not hire his songs out to bargains.

Music modern, now a thing of art,
    Fine art of difficult execution,
Introduced to meet new demands,
    Of this progressive age of fashion.

E'en that good old time Virginia reel,
    Of all the dances most inspiring,
The real test of the heel and toe,
    Condemned as fogy, undesiring.

'Tis true we hail from a fogy day,
    Since then the era of invention;
We were a happier people then
    In many ways that we might mention.

Those times never heard of matches,
    Lucifer (not Adam and Eve)
Lighted our fires with flint and steel,
    'Tis the truth, tho' it's hard to believe.

Our old time guns had priming pans,
    'Twas before the age of percussion,
Such the backwardness of the times,
    This we'll yield without discussion.

No steel pens nor gold diamond points,
    With goose quills all our letters wrote,
The coach or wagon our carriers then,
    Our freight often came by pole-boat.

Mails were slow and postage dear,
    Sealed letters with wafers or with wax,
Had no envelopes, no postage stamps,
    We write no fiction, but naked facts.

There was not a railroad in the land,
    Never a steamer plowed the sea,
Telegraph, telephone, all unknown,
    Nor dreamed of in all eternity.

Washing and churning done by hand,
    Pine torch or candles gave us light,
Nor stove, nor range to cook our food,
    Swinging pot-rack was our delight.

No thought of a sewing machine,
    Nor phonograph nor velocipede,
We had hearn tell of the elephant,
    In fact one of them we had seed.

Then the printing press was very crude,
    And pictures they were powerful scace;
Were way behind in all these things,
    But we were a mighty happy race.

Before diskivery of coal and ile,
    We still wonder at the electric light;
Now the street car, dummies they surprise,
    Reckon we was sorter in the night.

What comes next? We may learn to fly,
    And gold will be made out of clay,
Everybody will become so rich
    Will be nothing to do but play.

And then what next is hard to tell,
    May be that the girls will sprout wings,
Already they seem to fly around,
    And do some mighty funny things.

A word about the ladies of the day,
    One thing doth most sensibly impress,
Their skirts have got the natural shape,
    Resemble our old mother's dress.

Our fogy ladies were not so fast,
  They were too true and kind to flirt.
Did not go promenading round,
  But they kivered a heap more dirt.

Our kids had more respect for age,
  No, they warn't nothing like so peart,
If skedaddled round like to-day,
  Why, they'd hat-ter haul off their shirt.

Nor they didn't smoke the cigarette,
  But did inginerally chaw tobacker;
Did many things they oughten to,
  Our old-time fogy country cracker,

The female bustle was then well known,
  But was built in a different form,
Made out of rags and stuffed with tow,
  Or paper 'bout the size of your arm.

Prior to the days of hoopskirts, too,
  They warn't so waspish in the waist,
The dear sweet things didn't like to sting,
  'Twas before they acquired that taste.

There was a creature, that then unknown,
  Malformation now called the dude,
A hypersarcostic sort of thing.
  With little common sense imbued.

The changes wrought in fifty years,
  What we have told is but an inkle,
Invention is still upon the tare,
  Would stun a new Rip Van Winkle.

Now these public schools, the great jehu!
  They dish out larnin' by the platter,
Why, they pour it down and rub it in,
  They fry it in the children's batter.

Babies now are chocked and crammed,
  Are just loaded down with knowledge,
Soon as their bibbs are taken off,
  Are prepared to enter college.

George Washington was a mighty man,
  In days of which we've been speaking;
Could he come back, a goose he'd be,
  It would make him feel real sneaking.

Then what about old man Franklin?
  The man who first cotch the lightning;
Could he see half that's now been done,
  Why, he couldn't keep from frightning.

And that old-time orator, Patrick Henry,
  That liberty speech he seemed a cordon,
But, goodness gracious, in this great day,
  Whar'd he be 'ginst Grady and Gordon?

John Wesley, he was a "hustler" then,
  Counted powerful in a scrimmage,
But to set him down in these big times,
  With sich as Sam Jones and Talmage.

Old man Girrard was then thought rich,
  And so was Mr. John Jacob Astor,
But millionaires of the present day,
  Have possessions greatly vaster.

Merchants, lawyers, doctors all progress,
  Mechanics is the greatest wonder,
And farmers who held the biggest cards,
  Are the worst of all snowed under.

We seem to live in a different world,
  The old ball seems turned inside out,
Things aint running as they use to was,
  No, they aint agwine as they mout.

The sun's the same, the world has changed,
    Sun looks all right, shines as bright,
And the stars twinkle as they ever did,
    But the world has changed its plight.

The firmaments stand as firmly fixed,
    Our mother's Bible reads as of old,
The change must be in our fellow-man,
    He's' patterned in a different mould.

It may be all right, it may be wrong,
    The change may be to man's interest,
But we fear the devil's in the deal,
    For his cards seem to show up the best.

54 THE FOGY DAYS AND NOW;

FARMERS' HALL, PENDLETON, S. C.

Old Pendleton—A Sketch of Old South Carolina.

WHAT hallowed associations does the name of this old village conjure up—how often in thought do we wander back there. Old landmarks and many reminders are still to be seen, but the kindly faces and precious souls have nearly all gone across the bourne. We hope to meet them again in the better land, and if admitted into the eternal realms of bliss, and as time rolls on her endless cycles, we feel that now and then we should still be constrained to spare a moment to peep down upon the old familiar spot, where our first fond hopes on earth aspired and indulged in many bright anticipations, which have never been realized.

Fifty years ago old Pendleton was the fairest town in upper South Carolina, a community of wealth, intelligence, refinement and religion, and the home of the best people it has ever fallen to our lot to know. A resort of giant minds who would do honor to any age of the world's history—such men as John C. Calhoun, Langdon Chevis, Daniel Huger, Warren R. Davis, John Taylor, David K. Hamilton, the Pinkneys, Haynes, Earles, the Generals Pickens, Anderson, Blassengame; the Colonels Warren, Allston and Boul'on, and the homes of Barnard E. Bee, the Stevens brothers, of Charleston gunboat fame, of Confederate times, home of John and Pat Calhoun, the well-known young financiers of to-day; and from those old hills came our astute ex-Senator Joseph E. Brown, and

Atlanta's brainiest man, Dr. H. V. M. Miller; General Rusk, of Texas, a power in his day; Governors Perry and Orr, Commodore Stribling, of the navy, and hundreds who have left their impress upon this new world, and in their day and time helped to lay the foundation and build up this great country, and a host of others whose honorable names and useful citizenship would challenge the world for comparison. Such was the status of old Pendleton fifty years ago, when in the full tide of her prosperity. A splendid Piedmont climate, with fertile lands, and under the old slave regime; and then the wealth resided in the country, and agricultural pursuits were regarded second to none other as an occupation of honor and profit, and were conducted with an intelligence and advancement scarcely surpassed to-day in the South.

It was in the streets of old Pendleton that her indignant citizens kindled the bonfire that consumed in its flames the first incendiary papers and letters sent South by the abolitionists to stir up strife and discord among a happy people.

One of the first female high schools in the South was conducted there by Misses Bates and Billings, from Vermont, who taught the young ladies etiquette and French, graceful attitudes, and "highfalutin' notions," modern manners, to walk daintily, and to scream fashionably at a bug or a mouse.

One of the first military academies, where the boys drilled daily, and wore gray uniforms and brass buttons, was conducted there.

My first recollection of a Sunday School was there in the old Baptist Church, which is still standing. Uncle Tommy Sloan and Mrs. Fanny Mayse were the managing and leading spirits. We had little thumb catechisms, and the first and second questions were, "Who made man?" "Of what did God make man?"

The first cooking stove I ever heard of, my father bought, and was describing its excellences to Uncle Tommy, and among its other advantages he said: "Why, Tommy, it will save half the fuel;" when Uncle Tommy replied: "Well, Billy, why not get two of them, and save all the fuel?"

One of the first cotton factories was established at Pendleton and run with great success and profit for many years, and up to his death, by Major B. F. Sloan, and is still in operation by the Sittons.

Pendleton had her agricultural society, fair grounds and race track, and some of her exhibitions would put to blush many fairs of the present day.

Pendleton had four flourishing churches, two hotels; and who of her old citizens do not remember the long ball-room in the old Tom Cherry Hotel, and the beautiful young girls who once skimmed like swallows over those well-waxed floors, and the stately matrons, who, as chaperones, patronized with their presence these delightful occasions, and gave dignity and respectability to the ball-room? The old debating society, held in the old Farmers' Hall, and ever graced by a full attendance of the fair sex? The magnificent coaches and the elegant spans of horses that whirled up the dust in the streets of the old town? What old citizen's heart is not made to throb at the recollection of thrilling notes from the stage horn, borne over the hills to notify them of its coming? How the people would gather around the hotels and the postoffice as the great rocking, ponderous vehicle came rolling and swaying over the rocks, drawn by four or six horses, dashing in at a gallop into the center of the old town, with its passengers and mail. And with what eager excitement the citizens sought to welcome friends and visitors, and receive the tardy news.

Who does not remember the old "Pendleton Messenger" and Dr. F. W. Symmes, its editor, and the old "Farmer and Planter," and Major George Seabourne, proprietor and publisher; Mr. E. B. Benson, the long-time merchant, and old Billy Hubbard, the jolly landlord; the old English dancing master, Walon; rich old Sam Maverick, the eccentric; old man Sid Cherry, the bachelor; old Tommy Christian, the town marshal, and many other notables we have not space here to mention?

The first farmers' society in the South was inaugurated at old Pendleton in the year 1815, and was known as the "Pendleton Farmers' Society," and, if we are not misinformed, the second society of its kind in the United States, and the third in Charleston, in 1818, the first being in Philadelphia. The first officers of the "Pendleton Farmers' Society" were James C. Griffin, President; Josiah Golliard, Vice-President; Colonel Robert Anderson, Secretary; Joseph V. Shanklin, Treasurer and Corresponding Secretary. Its honorary members were General Thomas Pinkney, Honorable Wm. Lowndes, Honorable C. C. Pinkney, R. S. Izzard, Esq., J. R. Pringle, Esq., Doctor J. Noble, General Daniel Huger, Honorable John C. Calhoun, Colonel J. Boul'on, Colonel L. J. Allston, Reverend Doctor Waddell, General John Blassengame, D. P. Hillhouse, Doctor Isaac Auld, Doctor C. M. Reese, of Philadelphia.

And among the earliest resident members were Thos. Pinkney, John L. North, Andrew Pickens, Benjamin Smith, John Miller, Charles Galliard, John E. Calhoun, J. Taliaferro Lewis, Doctor Thomas L. Dart, General J. B. Earle, William Hunter, Benjamin Dupree, Joseph Gresham, L. McGregor, Samuel Earle, Richard Harrison, Patrick Norris, J. C. Kilpatrick, Joseph Earle, T. W. Farrar, C. W. Miller, Samuel Cherry, John Taylor, J. C. Griffin, Colonel Robert Anderson, Thomas Stribling, John Greene, Josiah Galliard, Francis Burt, John Hunter, W. S. Adair, William Taylor, William Anderson, Thomas M. Sloan, Joseph Mitchell, Thomas Lorton, Reverend James Hillhouse, Benjamin Dickson, Richard Lewis, J. B. Hammond, John Holbert, Robert Lemon, John Hall, David Cherry, Chas. Story, McKenzie Collins, George Taylor, Theodore Galliard, Samuel Gassaway, R. A. Maxwell, Jesse P. Lewis, Doctor F. W. Symmes, George Reese, James Farris, James O. Lewis, Henry McReary, David K. Hamilton, Major George Seaborn, Major R. F. Simpson, E. B. Benson, B. F. Perry, Geo. Reese, George Liddell, David Sloan, J. B. Perry, John Martin, T. Farrar, Warren R. Davis, William Gaston, John Maxwell, William Sloan, William Hubbard, Elam Sharpe, Leonard Simpson, Samuel Taylor, Major Lewis, William Steele, James Lawrence.

And this old Farmers' Society, organized seventy-five years ago, is still in existence, and flourishing under the administration of the present officers, D. K. Norris, President; J. C. Stribling, Vice-President; G. E. Taylor, Secretary and Treasurer; J. B. Sitton, J. D. Smith, James Hunter, W. H. D. Galliard, H. S. Trescot, Executive Committee.

Let all honor be given to the old Pendleton Farmers' Society, the pioneer of our Southern agriculture, the first organization of its kind in the sunny South, and nowhere in the State to-day can be found greener pastures, finer stock, or better farming, than in the vicinity of this venerable old village, all due to the grand race of people who once lived and flourished there, and at that time one of the most intelligent and delightful communities that ever existed on this checkered earth, and where to-day can be found a brighter galaxy of names and more honorable men than these recorded on the roll of the Pendleton Farmers' Society.

Once more, I say, let it be remembered in this ascending farmers' era, that from this little leaven came the leaven that shall leaven the whole lump.

There, too, was published one of the first agricultural monthlies in the South, under the proprietorship and management of Major George Seabourne, "The Farmer and Planter," a most able and valuable ally to the Farmers' Society, and did much to promote the spirit of agriculture in that section in its day.

It is the opinion of many persons now living that the author of the Junius Letters, so famous in their day, was a Pendletonian, one John Miller, formerly the King's printer, in London, and who fled from England on account of some political offense, settled at Pendleton, and was one of the founders and proprietors of the "Pendleton Messenger," seventy-five years ago. As far back as I can remember, the authorship of those letters were currently attributed to him.

But the glory of the old town has long since departed—in the first place shorn of her Samson locks, robbed of her territory and capitoley, the great district cut up into Anderson, Pickens and Oconee; and the railroads, of which she little dreamed then, have ignored her claims, stolen away her thrift, and now the good old town of auld-lang-syne stands out forlorn, gray and dilapidated in her tottering senility. But there still lingers a fragrance of intelligence and refinement in her social atmosphere that ever strikes the visitor with admiration and respect.

Since the days of which we have been speaking, the second and third generations are passing from the stage of action, rapidly losing their grip on life, and falling off into the sea of time. Of the second, Colonel Tom Pickens, Mr. Dickson and John Sitton alone remain, Mr. William Galliard having died but recently, and but a remnant of the third generation is left. The Clemson Agricultural College is now being erected at old Fort Hill, the John C. Calhoun place; a fine hotel is about to be built at old Pendleton, and it is thought the old town is looking up somewhat. May the Lord bless the faithful old spot, and may she become once more as she was in the days of yore, as a "city set upon a hill."

JOHN CALDWELL CALHOUN.

WITH all her honors in the olden days, perhaps nothing gave more distinction to old Pendleton than the name of John C. Calhoun, for that was his home. There he done his trading; there he schooled his children; there he and his family went to church; there he received his bulky mails; there many strangers came to visit him, and four miles from the town was his famous Fort Hill farm, a splendid property on the Seneca river, with broad acres of bottoms, fertile uplands and forests of native timber. The old home is still standing, a roomy but unpretentious looking mansion, overlooking the Seneca Valley and in full view of the Blue Ridge Mountains. This valuable estate was inherited by Mr. Clemson, Mr. Calhoun's son-in-law, and by him donated to the State of South Carolina for the purpose of an agricultural college, which is now being erected near the old mansion, which is, I understand, to be preserved intact, with the old furniture and bric-a-brac, that visitors may see the old home as it was in olden times.

Mr. Calhoun was very fond of his Fort Hill farm, and during his vacations from Washington gave much attention to his farming interests. He was first to introduce into that section blooded cattle, and I can remember his importation of the English red Devon cows. He first introduced Bermuda grass for grazing purposes. This grass is still to be seen on the great lawn in front of the old mansion, and I understand

JOHN C. CALHOUN.

this same Bermuda grass has about captured all of the fine bottom land on the place. He also first introduced the hillside ditches. I remember when I was quite a boy, seeing him superintending, surveying and staking off these graded ditches, and many times have I seen him with his eldest daughter, Miss Anna Mariah, walking together through the fields and meadows of Fort Hill.

Mr. Calhoun was ever pleased to receive and entertain his neighbor farmers and discuss with them the agricultural interests of the country, and it made no difference whether they wore broad cloth or homespun geans, all received the same kindness and attention. His most earnest friends were his nearest neighbors, and those who were best acquainted with his spotless character. No state ever held more confidence in her representative than did South Carolina in Mr. Calhoun, nor did the South ever have a better and truer friend ; he seemed to possess, to an eminent degree, all of the elements that belong to true human greatness ; though brilliant and profound beyond other men of his day, he was simple and unpretentious in manner, affable and conservative, yet as firm as the rocks of Gibraltar in his convictions ; possessed of a Christian spirit, without a shade of fanaticism, fully temperate, though not a total abstinent, gentle and kind in disposition, but with the heart of a lion when aroused by acts of aggression and injustice ; as to the depth of his great mind there seemed no bottom and his foresight of coming events is still the subject of remark and wonder to the present day.

John Caldwell Calhoun was born in Abbeville District, South Carolina, in March, 1792 ; his family were Irish on both sides. His father, Patrick, was born in Donnegal, Ireland, and landed with his parents in Pennsylvania when but a child. The family then moved to Virginia and from there to South Carolina, in 1776, where John, the last but one, was borne, and grew up on a farm; aspiring to an education he was sent over to Georgia to his uncle, Dr. Waddle, then a famous teacher and Presbyterian minister, and making such promising progress was next sent to Yale College, where he graduated with great distinction, and where, by invitation from Dr. Dwight, the president of the college, engaged with that celebrated scholar in a political discussion (they entertaining opposite views) which elicited from the doctor the remark: "That young man has talent enough to be President of the United States;" and the doctor predicted someday, that he would be, if he lived. From Yale he went to Litchfield, Conn., and attended the celebrate law school under Judge Reeves; returning to South Carolina he spent some time in the law office of Mr. Dessaussure, in Charleston, and also in Abbeville, S. C., with Col. Geo. Bowie. He served two sessions in the South Carolina Legislature and then was elected to Congress, and then to the United States Senate, and was afterwards Secretary of War, Vice-President with Andrew Jackson, and died as Senator from his old State, South Carolina.

Mr. Calhoun married his cousin, Floride Calhoun, of Abbeville, S. C., and settled permanently near old Pendleton, at Fort Hill. He raised seven children, Andrew P., Anna Mariah, Patrick, John C., James E., Cornelia and William Lowndes. Not one of the family are living, the two eldest being the last to die.

The eldest married a daughter of Gen. Duff Green, a man of great distinction in his day, and although Andrew P. was educated for and would have preferred a political life, was compelled to abandon the idea on account of his own and his

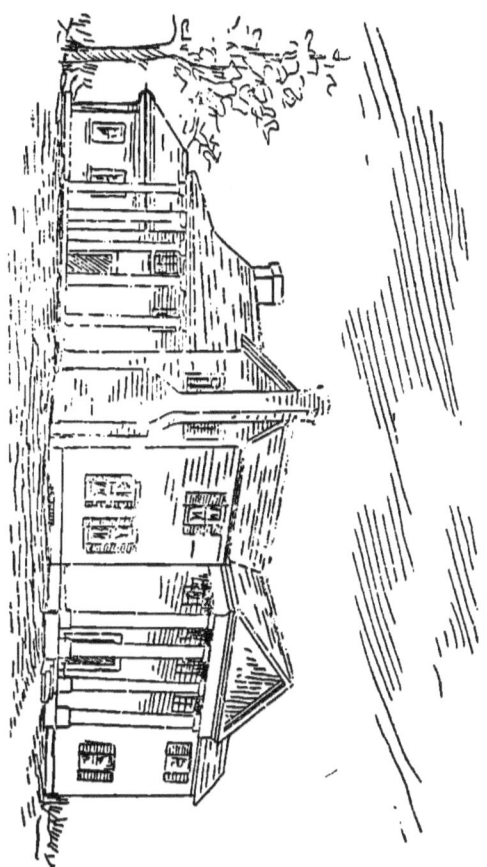
FORT HILL RESIDENCE.

fathers agricultural interests in Alabama, thus became a planter, and a successful one. After the death of his father he returned to the old home at Fort Hill, where his younger children were born. Anna Mariah, the next, married Mr. Thomas G. Clemson, a foreigner, and who was afterward made Minister to Belgium and Cuba, from this Government. He and his wife both died at Fort Hill, since the war. Patrick was a graduate of West Point, and died a United States officer, before the war. John C. chose the medical profession and graduated, but never practiced; he married twice and left several children. James E, perhaps the brightest mind of the family, settled in San Francisco, Cal., started out with brilliant prospects, but died quite young. Miss Cornelia was injured by a fall in her infancy and never married; she was also very bright, and assisted her father as his amanuensis. William Lowndes, the youngest of all the children, was my class-mate and best friend, married a Miss Cloud, of Winnesborro, S. C., and died early.

Mrs. John C. Calhoun was famous for her hospitalities and her varied domestic accomplishments; superintended, in person, her extensive household affairs; her home was ever full of visitors; she was the very perfection in housekeeping, and after the war her old house servants were in great demand; she was the most loving and indulgent of mothers, was very fond of building, and constantly kept carpenters in her employ, adding, changing and remodeling 'till the old Fort Hill mansion became a model for its conveniences.

Mrs. Andrew P. Calhoun is still living, and she and her only daughter reside with her son Pat, in North Atlanta. The family, like many other wealthy Southerners, were bereft of their fortune during the Confederate war, but through the phenomenal success of her sons, John and Pat, they are in affluence again. It is but a decade since these two enterprising young men were struggling for a support, and now have not only acquired fortunes, but have developed into great railroad factors, and have had much to do with the present prospective great development of the South. It is an undeniable fact that the R. & D. R. R. is largely indebted to the brain of young Pat Calhoun for its vast proportions. He is now a director and general attorney for this powerful syndicate, and John is also a director and is president of the Southern Society in the city of New York. Mrs. Calhoun has reason to feel proud of her children, and they are descended from distinguished ancestry on both sides. We doubt if there is a young man on this continent, to-day, who has accomplished through the means of his own brain, more than Pat Calhoun; but little past thirty years of age, he has climbed within the past eight years from absolute poverty to the top of one of the greatest money powers in the land, wields an influence and handles fortunes in a manner that savours of the old stories we read in the Arabian Knights. I feel I cannot close this reference to the family of Mr. Andrew P. Calhoun, without a word about Miss Margie, the only daughter, and do so without permission, and take the liberty, because I believe the name of John C. Calhoun and his discendants belong to the Southern public. Though Miss Margie's efforts have been confined to a more secluded field of action, she has proven herself no idler; she seems to have inherited her grandfathers taste for agriculture and fine stock, and in the management of her valuable stock farm near Atlanta she has exhibited a successful and practical business management that challenges competition with the best farmers of the day. Her management has been a success, as her green meadows and beautiful Jerseys and Ayershires will attest. She ignores cotton altogether, and confines her efforts entirely to forage crops, delights in fine stock, and sits on her horse as securely and handles the lines behind her spirited bays with the confidence of an expert. In many ways she reminds me of her grandfather, not alone in her predelections to agriculture and fine stock, but in the unassuming simplicity of her manner, and her disposition and capacity to entertain and interest others.

In the dark days of the Confederacy, and after the war, during the years of poverty of the widowed mother, this daughter became the great comfort to the mother and the sole instructress to her younger brothers, and besides her teaching, Pat went but a short time to Prof. Cooledge, at the Dalton Academy, and to the devotion and inspiration from this noble sister he is, no doubt, indebted in a large measure for his extraordinary success in life. Miss Margie seems to have no aspirations for herself; her whole ambition in life is concentrated in the interests of her brothers; she prefers the most simple and retired life and is chief and major domoress of the entire home department.

I love to think and talk about the John C. Calhoun family, they were our nearest neighbors and best friends; the Calhoun boys were my school mates for years, and they were my associates all through the days of my youth—we rode to school together to old Pendleton, hunted together in our holidays and the youngest son, William Lowndes, was my bosom friend; his mother used to call us her "Damon and Pythias." The first enterprise I ever attempted, he was my partner. I have outlived them all, and shall ever cherish in the greenest spot in my heart every member of that noble family.

I have often conversed with Mr. John C. Calhoun, for he was fond of talking with boys, and would adapt his conversation to entertain and instruct them. He once said to me, "You boys go out hunting with your double-barrel guns, powder flasks and shot pouches filled with ammunition, and not even the little larks and bullbats escaped your attention, you waste your ammunition and bring home trifling game in your bird bags; said it was not so in his youth, that then he shot a rifle, and never fired at anything less than a squirrel or a turkey, and that it was a rare thing for him to miss a shot; that amunition was expensive and had to be economized. I still have in my possession his life and speeches, presented me by his own hand. He also gave me a list of histories for my early readings, which I purchased and kept up to the late war, and lost during the confusion of that terrible time, together with everything else I owned. Mr. Calhoun told me that his favorite reading in his youth was such books as Josephus Rollins, Ancient History and Plutarchs lives, and especially the last, he was very fond of; said we boys were too fond of trashy novels, that he never read them. I remember once discussing with Mr. Calhoun the phenomena of rains, his unassuming manner throwning me off my guard, when I launched off into quite a theory of my own. He listened deferentially to what I had to say, and then gave me modestly his ideas upon the subject, and I was so struck by his able logic that it suddenly occurred to me that I was listening to the greatest mind of the day. How ridiculous my shallow ideas must have appeared to him, and during the balance of the conversation felt constrained to say little more than yes sir, and no sir, and felt much embarrassed, which I know he discovered and tried to relieve.

Traveling through the Blue Ridge Mountains in North Carolina, Mr. Calhoun, Col. Gadsden and my father stopped over night at a mountain cabin home. There was but one spare room, and in it a bed and a pallet. My father arranged for himself and Col. Gadsden to take the pallet and Mr. Calhoun to take the bed. About midnight the mail-rider stopped in, and seeing but one person in the bed, said: "Git furder thar, old horse, and spoon," and familiarly piled in with the Senator. In the morning the hostess came in the room and finding Mr. Calhoun there alone requested him to climb up a ladder into the loft, and hand her down a shoulder of bacon, which the Senator complied with, as gracefully as circumstances would permit.

Our party spent several days on this trip in Cashier's Valley, at the home of the old man, James McKinney. Mrs. McKinney was quite a stout, red-faced, middle-aged lady, celebrated far and wide for her curiosity as well as her loquacity, as also her unsophisticated manner; entering the room where the gentlemen were talking, with her sleeves rolled up above her elbows, her arms akimbo, addressing my father, with whom she was acquainted, said : " Colonel Sloan, is this the great John C. *Cal*-houn that I have hearn so much talk about ? " My father answered in the affirmative, saying: " Mr. Calhoun, allow me to present to you our hostess, Mrs McKinney." Mrs. McKinney grasped the proffered hand, saying : " Do tell; why, you look jist like other folks. I reckon you've got a mighty purty wife to home haint ye ? " Mr. Calhoun answered, that he intended bringing Mrs. Calhoun on a visit to the mountains, and she would have an opportunity to judge for herself, when Mrs. McKinney broke in again, " Well, I low she's got lots of purty bed quilts down thar," when old man McKinney spoke out, "Thar now, Sally, you've played h—l agin," and for one time in his life our great Statesman seemed at a loss for a reply. Mr. Calhoun made frequent visits to these mountains with my father, examining the topography of the country in view of a railroad crossing the Blue Ridge, and could often be seen cracking rocks in search of minerals. He was first to discover the indications of gold in that section, and afterward, my father and others, worked extensive gold mines there.

Mr. Calhoun was noted for his wonderful forecast of coming events. Many are still living who remember his predictions about Marthasville, now Atlanta, the coming city of the South. Nearly fifty years ago he said it would become a great railroad distributing point and a great city. He greatly desired about that time a railroad connection between Charleston, S. C., and Knoxville, Tenn., which enterprise was finally undertaken before the war, and after an expenditure of several millions of dollars, under bad management, was abandoned for want of further means, the failure proving a great misfortune to South Carolina.

As a boy, I have often heard Mr. Calhoun discuss with my father the great approaching crash between the North and South, and its certain fearful results. He would show the continual encroachments of the Abolitionists upon the constitutional rights of the South, and pictured the troubles that would be unavoidable. He feared that our people did not fully appreciate the gravity of the situation. I have often heard him say it must come if these aggressions continued, and from his intimate knowledge of our opponents and their unrelenting and selfish character, he feared the worst.

My father was devoted to Mr. Calhoun, and when he died was in the deepest grief and gloom, for he felt that the greatest, the wisest, the purest of all, was lost to his country; and I cannot help but believe that could he have lived untill our trouble, we would have come out of it better than we did.

I remember once that as father and I were riding over to Pendleton, passing the Fort Hill big gate, we discovered Mr. Calhoun and his negroes fighting fire in the woods. We got out of our buggy and assisted in putting out the fire and saving the fencing.

When the Senator would return from Washington, my father and other neighbors would visit him frequently, being received at his library, a cosy little house out in the yard under the shade of several venerable oaks, where they would discuss the state of the country, agricultural, and other topics of the day.

I once encountered a most embarrassing position at the Fort Hill dinner table; had been out hunting with the boys and returned to dine with them. Mr. and Mrs. Calhoun were seated at either end of the long dining table and I was placed right between two very elegantly dressed young ladies (suppose everybody has some peculiar weakness, and somehow, elegantly dressed young ladies, with long trains, always had a paralyzing effect upon my mental system), my embarrassment increasing with the closer contact. I was feeling exceedingly awkward and cramped on this occasion, when to increase my discomfiture, my friend, Willie Calhoun, requested me to carve a roast duck just in my front. I picked up the carver and fork and made an awkward lunge at the fowl, when it skipped clean out of the dish, landing plump into Miss Martha Calhoun's lap. It was an awful affair, and my first impulse was to fly, but I dared not attempt it, for those long, mysterious silken trains were coiled all around about my feet, and I feared if I made a dash I might become entangled in these stylish appendages and upset the young ladies or something else, and the only resort I could think of was like old Adam, to try and put the trouble off on some one else, so turning to the sufferer I said: "Miss Martha, I am very sorry, this thing would never have happened if it hadn't have been a wild duck." This excuse brought down the house with a roar of laughter, and even the stately Senator smiled and remarked that the young man should be pardoned at once, and the pardon was at once graciously granted, and instead of the miserable culprit, I at once became the hero of the occasion.

I now want to tell you a story. It may at first appear a little marvelous, but I have earnestly tried to give the truth all through this little book, and as I am now a white haired man, am persuaded that I have borne a respectable name for veracity, and would therefore regret, at this late date, to be considered a competitor of the Baron Mon Chaussen. I trust the reader will at least be kind enough to give the statement the benefit of what the law recognizes as ' reasonable doubts." The strange story is about a remarkable and very deep old well on the top of old Fort Hill, on the John Calhoun place.

It is said that, in olden times, several battles were fought around this old fort, and reported that many human bodies were thrown into the old well. It has never been used since. In our day, there was much superstition about this old well, especially among the negroes, who gave the place a wide berth after nightfall, but as to the facts of which we are about to state, I, and others now living, were personal witnesses.

In that day, if a person would go to this old well after sun set, and leaning over so as to throw the voice down to the bottom of the well and would halloo, "what are you doing down there?" It would answer back, "n-o-t-h-i-n-g a-t a-l-l." As to the whys and wherefores, we decline even to express an opinion, but leave our incredulous readers to form their own conclusions; we can only avow again that we have in no wise misrepresented the facts.

Mr. John Ewing Calhoun, a brother of Mrs. John C. Calhoun, married the sister of the distinguished South Carolina Congressman, Warren R. Davis, and owned and lived on a splendid estate of lands adjoining Fort Hill; had also many slaves, and was considered a very rich man in that day. It was his son, Col. Ransom Calhoun, who was killed in a duel by Lieutenant Rhett, on an island near Charleston, in the early part of the war. His only daughter, Miss Martha, familliarly called Coody, was one of the most splendid young ladies of that day, with a cultivated intellect, a gifted conversationalist, an accomplished musician and the author of the "Keowee Waltzes," and besides an equestrian of extraordinary skill. I once saw her mount a young blooded horse of her fathers" that two negroes with difficulty held whilst she was being seated, and when turned loose, skilfully managed him. She died early—never married.

Mrs. Calhoun had another brother, Mr. James Edward Calhoun, a very wealthy man, and one of many eccentricities, who lived on the Savannah River, in Abbeville Dist., S. C., and who died but recently at a very advanced age. He left no children.

One of her Sisters was the wife of Gov. Noble, of South Carolina. The Calhouns, of Atlanta, are also of the Abbeville, South Carolina, Calhoun family. We refer to the great oculist, Dr. Abner Calhoun, and Judge William Lowndes Calhoun, who has filled so many honorable and useful offices in this City; the brilliant lawyer called "Andy" Calhoun, and other members of these families.

Circumstances have connected a name with the John C. Calhoun family that savors much of romance, that of James H. Rion, late of Winsboro, S. C., who died a learned scholar and distinguished lawyer. Of all persons now living, I am, perhaps, the only one that can give a correct account of his connection with the Calhoun family, who had much to do with the shaping of the remarkable events of his after life.

How well I remember the first time I ever saw Jim Rion, sitting alone on the root of a great oak in front of the old Pendleton Academy, with a yellow ribbon band around his plain straw hat. I was struck with the peculiar whiteness of his skin, his delicate and girl-like appearance. I spoke to the boy and learned from him that he and his mother were Irish Canadians, but recently from Savannah, Ga. His mother had come to keep house at the Old Pendleton Hotel. He wanted to witness the examination then going on, but was too timid to venture in alone. I conducted the stranger boy in and shared with him my seat.

Soon after this he entered school, becoming my classmate and we afterward became devoted friends, he spending his Saturdays and vacations with me at my fathers beautiful home, Tranquilla, on the Seneca River, and in the Blue Ridge Mountains, hunting.

I rode to school at Pendleton, joining the Calhoun boys at the big gate. One morning, calling at the mansion Mrs. Calhoun mentioned to me that she wanted a good housekeeper, when I told her of Mrs. Rion, whose cakes and pies I had so often enjoyed, and at the request of Mrs. Calhoun I went to see Mrs. Rion and obtained her consent to go to Fort Hill, and then Jim formed a part of our cavalcade to the Pendleton Academy. Our party consisted of Mr. Calhoun's three sons, John C., James E., William Lowndes and Jim Rion, from Fort Hill, Ransome Calhoun, from Keowee, and my Uncle John Hackett and myself from Tranquilla.

Jim Rion received every kindness from the Calhoun family, and it is believed, to this day, even in South Carolina, that he was of blood relation to the Calhouns, but it is not true. Jim Rion was fifteen years old when he came to Pendleton, and sixteen when he went to Fort Hill.

My father first noted his brilliancy of intellect and spoke of it to Mr. Calhoun, and through his influence, and the efforts of Young James E. Calhoun, who was then in college, he was entered as a beneficiary and graduated with great distinction, winning the first honors of his class, though some of his competitors belonged to the wealthiest and most aristocratic families of the State. He also captured a more precious prize from the family of the President of the College, then the Hon. W. C. Preston.

Rion commenced business as a teacher in Winsboro, S. C., studied law under the famous Mr. Woodward, and soon became his partner in the practice. The war coming on, he was among the first to volunteer, came out as a colonel of a regiment, and was known as a brave and brilliant officer. After the war he rose rapidly in his profession and became famous as a railroad lawyer, which branch he made a specialty. He refused to enter politics and to accept any kind of political preferment. He presented two scholarships to his alma mater, in gratitude for benefits received. Mr. Calhoun entertained a very high opinion for James Rion, and in many ways showed his confidence in his talent and integrity, and after Mr. Calhoun's death, Rion found opportunity, and did render valuable service to members of his family. James Rion was a remarkable man, and his death was not only an irreparable loss to his family, but to his adopted State.

THIS DAY OF PROGRESSION.

The world moves on, it does progress,
    Rests not, rushing on, on it goes;
Where, or whitherward it may be bound,
    Is veiled—God himself only knows.

We may look back for fifty years,
    And our records tell of many more;
New petals bud and then unfold,
    And each one gives a greed for more.

To-day every man's for himself,
    Hindmost left to the devils care;
The tickling game, the winning card,
    Man must tickle, to get his share.

Friendship is but an empty sound,
    And gratitude a giddy farce;
Going up we meet many a friend,
    But coming down we find them scarce.

Who to-day is our dear neighbor,
    On whom are your praises lavished;
They who sit on the topmost rails,
    Favors wanted there are ravished.

Boot licking full in fashion now,
    A science made of flattery;
Success, hard won without deceit,
    The tickling force the battery.

Vile rings are formed to consumate,
  And foulest schemes are designed;
Oft worse men are placed in power,
  To swindle the weak are combined.

This day an humble, honest man,
  Is trodden under out of style;
Is tho't a man of no git up,
  Voted out of rank, rank and file.

'Twas in this progress we fell short,
  Sought for the truth, even prized it;
Had no more sense than pay our debts,
  Condemned trickery, despised it.

Wore woolen jeans, home tanned boots,
  Had shirts hitched to our collars;
Our breeches had the fogy flap,
  But pockets filled with dollars.

Yes, you call us old-time fogies,
  Our old-time ways you have dropped;
New things, new ideas every day,
  From the fogy world you've flopped.

If your progress was most for good,
  Good and evil both run along,
Side by side, do their waters flow,
  But evil seems the biggest prong.

One flows on with gentle ripple,
  Other rushes with a mighty roar;
The one, but laves its gentle banks,
  But the other is flooding o'er.

The first progression known was sin,
  First development brought unrest,
First advance was in devilment,
  And proved to be a bad invest.

First couple were a happy pair,
  Until they struck that progress tree,
A curse upon the first invention,
  An apron to hide naked-i-tee.

Proof first progress was not for good,
  In fact, panned out vice-versa,
First advance to pollute the soul.
  Forward movement, bad disburser.

May be all right, it suits the folks,
  Who are the makers of the times,
Progress is the new order now,
  And the propelling force, the dimes.

Our old time rig is of ante-date,
  Modern innovations now preside,
Be it for better or for worse,
  They are the court, they must decide.

New modern fleets around us tack,
  See gaudy yachts go flying by,
Fast steamers leave us in their wake,
  Cant keep up, 'taint no use to try.

Are on the sail, must scud along,
  Old bark must buffet with the tide,
Rough breakers beat against our prow,
  And the great sea seems drear and wide.

Are on the train, its schedule new,
  We can but wait, and watch, and see;
But it seems it's running very fast,
  Too fast, much too fast for we.

AN AGE OF MONOPOLY AND GREED.

An age of monopolous rings,
    Formed of sharpers, bulls and bears,
Of cunning trusts to rob the poor,
    Then divide up their guilty shares.

The rich grow richer every day,
    Their greedy craws never satisfied;
They grind the poor down into the dust,
    And e'en life's comforts are denied.

While millions live from hand to mouth,
    Are o'er burdened with the toils,
Monopolies' coffers are never filled,
    Nabobs are gloating in the spoils.

An age of swindles and humbugs,
    Honesty stands but little chance,
The big dog's got the whip in hand,
    And the little dog's got to dance.

Now these big dog's have got to think
    That they are made of porcelain clay,
The under dog's of common mud,
    That they have lost all right to say.

They look down on poverty as shame,
    Though it meet all its obligations;
A guilty shame that they would shun,
    As only fit for their abnegations.

And distress of odious savor,
    To some who recline on roses,
On earth they turn up their noses,
    Nor change till up go their toe-ses.

What a sight in the judgment day,
    When the Lord gives out his diplomas,
When some of these high-stepping bucks,
    Will skulk away with the gloamers.

When earthly laws shall be reversed,
    By right verdicts of the master,
When conflicting judgments of men,
    Be wrecked in common disaster.

Then ye high-headed ones of caste,
    Don't deign a nod to your betters,
What will become of your pewter?
    For you must hand in your letters.

What will you do in the awful day?
    When the grim monster shall find you,
The old imp, with sulphurous breath,
    Brings his icy chains to bind you.

We would not chide at all the rich,
    For the world must have its pageants,
For Heaven fills many a purse,
    And makes good men its agents.

Grand the man who is so blessed,
    With worldly wealth and with a soul,
A heart to feel and hand to help,
    For he shall reach the highest goal.

The biggest fool that we can ken,
    Though he be a man of learning,
The bloated toad who loves himself,
    His own great traits alone discerning.

THE FOGY DAYS AND NOW:

Sometimes we meet him on the streets,
   Have marked his braggart swagger,
Oe'r the humble, he towereth high,
   To such his eye hath look of dagger.

'Twould be hard if in this poor world,
   If recompense in this dark vale,
What entanglements wright and wrong,
   Thank God there is a grand finale.

What is progress, except from sin?
   What worth the earth, its passing joys?
A few short years when we look back,
   These mighty things will seem as toys.

The rage now is to let 'er roll,
   Roll on, rush on, regardless where,
Let 'er roll, we'll cross the stream,
   Though we know the maelstrome's near.

Sometimes we gaze into God's expanse,
   Peer out into a thousand years,
Then look back at the trifling past,
   And smile at former joys and fears.

See how we struggled there for naught,
   Some worthless bauble to obtain,
How many mistaken roads we took,
   And how suffered there in vain.

Then we laugh at human giant fools,
   Whose form once towered o'er the poor,
But pigmies do they now appear,
   Shivering dwarfs outside the door.

We see the once grand millionaire,
   Who had but borrowed deceitful gold,
That swelled his purse a little while,
   And then found too late he was sold.

Have some magnates now in our mind,
  Who in these great times do dwell,
Have lorded over God's little ones,
  Might call their names, don't care to tell.

Sometimes the case that rulers be,
  Whose hearts are rotten to the core,
Clothed in power for one brief day,
  And who may soon be worse than poor.

THE TWO STREAMS.

But there are good as well as bad,
　E'en in this wild and rattling day,
Heaven hath sentinels every age,
　To point its pilgrims on their way.

Midst sin and shame, some humble ones,
　Unknown, unselfish, every thought,
Whose secret prayers reach His throne,
　Who have his ardent battles fought.

There is a Christian type this day,
　Same as was in the days of old,
As high, as true, and noble too,
　That ever watches o'er his fold.

And these make up that gentle stream,
　The stream that laves its placid banks,
And but for these the world were lost,
　For these let's give to God our thanks.

To noisy world are often hid,
　Unpublished, all their work is done,
In self-denial hold their creed,
　And through faith is victory won.

Not always in the pulpit found,
　Not ever in the church would seek,
Not midst the gay and social realms,
　But rather mongst the low and meek.

The angels know them, man's in doubt,
   May be Lazarus or Magdalene,
Such that might be scorned of men,
   Such might be God's choice I ween.

How many wolves that wear lambs wool,
   Have gathered within the fold,
They may deceive his people here,
   But in his courts they cannot hold.

How often e'en within the church,
   That Heaven's temples are profaned,
Hypocrite in a deacon's chair,
   Who for a saint hath been ordained.

How many, who may feel secure,
   That will pass through the inner gate?
Oh! How many shall enter there?
   And how many will miscalculate?

Then I wonder what'll be my fate,
   When I'm called to make the change,
If I'm saved for what I've done,
   Would think it passing strange.

If I'm lost, could but deem it just,
   For I know I've a rebel been;
Could make no excuse, silent be,
   My sins I would not dare to screen,

In soul I know I love the Lord,
   But in the flesh I'm very weak,
Sometimes I feel a would-be saint,
   Then comes again a devlish streak.

Now some may think all this is weak,
And to all such it may be Greek,
To me the only solid plan,
Only reasonable left to man.

All others fail of which I've read,
To suit the living, fit the dead,
None like Christ on earth hath trod,
Born a man, I believe he's God.

I accept Him; in Him have faith,
His promise seal what e'er He saith.
Against His word dare not reason,
Simple, sacriligious treason.

I shall cling to his written word,
'Tis ahead of all I've ever heard;
If some things I can't understand,
Still I'm subject to his command.

Believe in both heaven and hell,
Where right with wrong can never dwell,
Deprived of heaven, all that's good,
Is essence of hell, its daily food.

EARTH'S THREE EPOCH'S.

The earth lay dormant, a dismal mass,
   Its first epoch thus for ages lay,
Ever whirling, turning, turning,
   For how long, God alone can say.

Next a wriggling of created worms,
   Then chattering noises in the air,
But in Eden sprang the master worm,
   Then a wormess made the happy pair.

But too soon they spun their first cocoon,
   With silken threads, their funeral shroud,
Which their fair forms was to entomb,
   Heretofore had never seen a cloud.

Human destiny ruined, one fell swoop,
   Thwarted, blasted by a devil's trap,
Banished hence from that garden's bliss,
   Forever by this sad, sad mishap.

Henceforth doomed to labor and to plod,
   Earth's second epoch, a faithless race,
Stupid fogies, seeking how to find,
   How to dodge the law, sweat of thy face.

And though cycles of time have they spent,
   Have striven and toiled to attain,
To save "elbow grease" have been intent,
   Striven and toiled, and still in vain.

The old cocoon they at last have burst,
　On aerial wings they now seek to fly,
As butterflies sail on gaudy wings,
　Flutter in the sunshine, then must die.

This century, fogy chains were loosed,
　And since invention can scarce be told,
And to-day they sail on gaudy wings,
　They do hardly seem the worms of old.

The old fogy worm was sleek and fat,
　And he was content his sphere to fill,
And these butterflies they flounder too,
　With all their gaud are but mortal still.

They gambol midst sweets of every kind,
　And reckless, no thought of coming storm,
Forget the tempest is sure to come,
　Beneath flower lies the prostrate form.

And such is life, then what doth it wot,
　In this brief life whether crawl or fly?
How short at best our troubled days,
　For in the midst of life then must die.

And the spark of life is all the same,
　Let outside be worm or butterfly,
This vital spark is all that's worth,
　The only part that can never die.

The vital spark alone can stand,
　For all else is nill, good for naught,
The God-given spark to every man,
　Only spark of earth from heaven caught.

Then what matter whether we crawl or fly?
　What matter whether we sail or plod,
Best of all to live an honest man,
　To be the best, the noblest work of God.

And through all times we now conclude,
　There have lived upright, honest men,
Not so many as there used to be,
　But still they do turn up now and then.

THE PEWTER BUCKLE MOULDS.

ONE bright morning my father sent me up to old man Howell's, with an order for a lot of shingles. The place was about six miles off, and I was soon on my way, galloping along the country roads. It did not take me long to reach my destination and learn that the old man was already out in the woods, drawing shingles; however, his son Mart was at home, and kindly offered to conduct me to his father, but before starting, exhibited to me an invention of his own, a pair of soapstone buckle moulds, also displaying a stock on hand of bright, shining gallows buckles. I examined the machine with undisguised wonder, feeling that I was in the presence of a genius, and looked upon the inventor with profound admiration. This poor man's son, without opportunities, with his untutored hands, had wrought this valuable machine. I thought, what a brilliant future would be his, what wealth and fame would fall to his lot in life; such were my meditations as I inspected the beautiful products of his invention. I was startled from my reveries by the young man proposing to sell the moulds to me. I had not imagined he would part with this valuable property for thousands. I was still more surprised when he offered to take the insignificant sum of $1.50. I had but 75 cents, but he took that rather than miss a trade, admitting he had sold too cheap; then said there was a soapstone quarry close by and he could make more moulds; that he was short of capital, and this sale would enable him to lay in another stock of pewter and go on with the business. He said the world had to be supplied with these gallas buckles; that there was a great future in the business for both of us; that my engaging in the business would only help to advertise it; that it would take a number of factories to supply the demand; that he was willing to share both the fortune and the fame with me.

I purchased the factory, and was so elated with my investment that I came very near forgetting the errand upon which my father sent me. It was not long before I was on my return home with the valuable machinery in my breeches pockets, engrossed in the contemplation of a great enterprise to be established in the very near future. That night I tossed restlessly on my pillow and couldn't sleep for pondering upon my great scheme. I organized many brilliant plans for the future operation, but determined to keep my counsel, for I had heard it said that a wise man keepeth his own counsel. In my good mothers kitchen I knew there were large numbers of pewter spoons and plates, all of which I determined to capture and convert into valuable articles of trade. I matured many important plans of procedure during that short night.

Next morning I arose early and made a confident of black Dan, my fathers hostler, who had often proved my faithful friend and allie when I wanted a horse out of the stable at night to ride fox hunting. With Dan, I held a protracted and secret caucus, and it was agreed to go on a 'possum hunt (to all intents and purposes), so after supper we tooted up the dogs and sneaked all the pewter out of the kitchen, secured an old bullet ladel for melting the metal, then repaired to a deep hollow not far away, built a fire and started the factory. Everything worked like a charm, the enterprise was a success, and we continued to mould gallus buckles until the first warning notes of the morning cock admonished us to desist. We had near a peck of the shining beauties on hand; it had proved a glorious triumph, and I and Dan were happy; we congratulated each other, shook hands time and time again. I promised to make Dan a foreman in the factory, and, in a few years to set him free, give him eighty acres and two mules.

I now determined soon to hold a conference with my parents, and thought what a surprise it would be to them; and oh, how happy it made me feel to think of their delight in the discovery of the enterprise and cleverness of their eldest son. I determined that very day to show up the whole scheme, together with my well digested plans for operation in the business. I intended to make my worthy sire a principal partner in the concern, and we should either employ young Howell, or give him an interest in the business. His department would be to make the moulds—make them on a grand scale. We would have double moulds, tripple moulds and after a while, an acre of moulds, and great cauldrons to melt the pewter. My father could make a corner on all the pewter in America, get an option on all the timber in the neighboring counties for fuel, get up all the labor possible, and when once under full headway, would run the business for all it was worth. We would become many times millionaires, would build churches, schools and hospitals, help the poor and afflicted, and in my great gratitude to a kind providence, I resolved that no one within my reach should hereafter suffer for want of good, remunerative labor, or the comforts of life, and I did not know but that under a favoring providence I might become an humble agent in the ushering in of the great millennium.

Early after breakfast, I took Dan and made a visit to the

buckle factory. I shall never forget the joy I felt that morning as I surveyed that embryo buckle factory, the pride of that epoch in my life's history, with what complacency and self confidence, with what intensity of satisfaction with myself and all the rest of mankind, as I stood there with my arms folded across my peaceful breast, contemplating the vast fortune that had so benificently fallen into my lap. Oh, could I have died right then; but alas for all human hopes, when we feel strong it is so often but the precursor to our own weakness. While thus wrapt in the glories and fulness of my great enterprise, black Dan was hitching on a pair of buckles to his home-made gallases, but the tongues to the buckles bent and easily broke; they wouldn't hold, and a sharp exclamation from the negro broke up my reveries. "Why! Marse Dave," he exclaimed, "dese buckles, dey aint no good, look-a-here." I saw it, the truth flashed upon me like a thunderbolt: it staggered me. I tremblingly asked, "what's that Dan?" He answered, "dese buckles." It was enough, I was stupefied, squelched; this was a part of the business that had been completely overlooked; ruined, bursted, at one fell swoop a bankrupt. There was only one case of equal gravity that I could think of, and that was when Lucifer fell from Heaven. Instead of the great millionaire, as I had calculated, I was a pauper; instead of one who had achieved both fortune and fame, I was now a miserable culprit, for I knew the pewter plates and spoons had to be accounted for. I turned from Dan, my faithful colleague, in gloomy silence, spiritless and hopeless, bearing my almost paralyzed body back to the parental mansion, where I found new troubles awaiting me. My mother had the cook up to answer for the missing utensils. I could not allow the poor innocent woman to suffer for me; I told my mother it was I who "cut the cherry tree with my little axe." I then laid open to my compassionate and sympathetic mother the whole story from beginning to end, as best I could, between sobs, and my kind and considerate mother concluded I had already been sufficiently punished, and even tried to console me under this, my great trial, but somehow the whole affair leaked out and became the talk for a full week in the neighborhood. I was greatly prostrated for a time, but finally recovered my wonted enthusiastic disposition.

Years after this occurrence, I visited Milledgeville, Ga.; went to see the penitentiary, and among the convicts I discovered my gallas-buckle mould inventor, Mart Howell, making shoes for the State. My talented friend had surprised me once more, and upon inquiry he explained that he had been unjustly incarcerated in that unhallowed place; said he was a martyr to cruel circumstantial evidence; that some years ago, while in attendance on a camp meeting, just for a joke, he took a fellows horse and rode a little ways out, intending to bring it right back, when a crowd of rascals got after him and accused him of wanting to steal the horse, when such a thought had never entered his head; said he was just about to turn around and go back with the fellows horse when they came upon him. I asked Mart how far he had got with the fellows horse when they came upon him? He answered, "But a little ways, not more than twelve or fifteen miles;" and such is life; our inventive genius in the Georgia penitentiary, and a would-be millionaire and philanthropist keeping an Atlanta boarding-house.

"I set me up a bakers' shop,
 And thought I was improving,
But a bakers' shop will never do,
 So must push along, keep moving."

MY FIRST HORSE TRADE.

From first remberance, I have been impressed with the idea that I was a born speculator, and have ever been in expectation of some grand result from this inherent talent, although my experiences in life have turned out to the contrary, I am still unshaken in my faith, and live in constant expectation of something turning up. I can only reconcile the past contradictions to this (my pet theory) by the belief that there is but one little obstruction in the way, and that is, as yet, I havn't happened to strike it right; have not struck the flood tide at its proper stage, and though I now number past three score years, still in the vigor of manhood, I have not dispaired. I feel my good time has got to come, and if I fail to catch up with it in this world, which I have now concluded is most probable, then I shall confidently expect to be successful in the next one.

My father owned and worked extensive gold mines in North Carolina, though our home was on the Seneca river, in South Carolina. He often sent me to Dahlonega, Ga., where a United States mint was located, to have the gold dust coined. I started out one beautiful spring morning with some two thousand pennyweights of this precious stuff in my saddle-bags, riding a splendid young sorrel mare named Francis; my father and mother both stood out on the portico and watched me as the beautiful filly bore me gracefully from their sight.

Proceeding on my journey, about noon, I overtook a small, stoop-shouldered, freckled-faced, red-haired man, about forty years of age, wearing a wool hat, blue jeans coat, coperas breeches and home-tanned shoes, without socks. He was jogging along in a sort of lazy mixed walk and trot, on a sluggish looking old sway-back, clay-bank mare, with flax main and tail. Feeling lonely I accommodated my gait to suit that of the stranger, and we soon became engaged in conversation. Several times I noticed the man eyeing my handsome filly, and after awhile he ventured to remark: "That's a right snug critter you've got there, how'd you like to swap her?" Swap for what?" I asked, astonished at his impudence, "you don't mean for that old thing you are riding there do you, why, I wouldn't have her as a gift?" He mildly replied that he was not at all surprised at my hastily formed conclusion; that he took no offence at what I had said; that it would not always do to judge by appearances; that his critter was calculated to deceive more experienced heads than mine; said some of the most famous horses in the world were the most unsightly looking; said his critter had royal blood coursing through her veins; called my attention to her pointed ears, wide nostrils, the full swelling veins, the symetry of her limbs; said she was now with foal by the celebrated horse Steel (a horse that I had seen, and the most famous horse of that day) and that the colt would bring five hundred dollars when it was six months old. I listened with wonder at all this rig-a-ma-role, somewhat staggered as he talked on, but still unconvinced; after awhile I ventured to make an objection to the color of the mare. He quickly replied that he was glad I spoke of that, as her color was one of the best evidences of her value, and asked me if I had not, myself, observed that all circus horses were selected for intelligence, and that the white and spotted, and especially the clay-banks for ring purposes; this last claim for the old mare was a stunner.

He saw he had gained a point; said he didn't want to part with the critter, not for love nor money, and wouldn't think of it, exceptin' for the fact that he was now on his way to the Mas-se-sip, and owin' to his critters condition he was afraid he would hatter leave her somewhar on the road, and as he had taken a considerable liking to me, if he was obleeged to give her up, would ruther put a good trade into my hands than in the hands of a man he'd never seen before.

My opinions had now undergone a complete change in regard to both the little man and the old mare. The kind expression on the little man's face made him seem to me now real handsome, and the entire aspect of the old mare had changed; and although at first she had excited my disgust, and what I considered deformities, were now points to be estimated. This same old, ungainly animal, had become the great object of my desires. I observed closely, and in great admiration, the pointed ears, the wide nostrils, the swelling veins, and magnified the royal blood coursing through the intelligent animals veins. At last I asked my new found friend how he would be willing to trade? He answered reluctantly, that it made him sick to think about trading the critter off, but it seemed he was obleeged to do it; said to come to the real worth of his critter, he couldn't expect to get nothing like it, that there ought to be a great deal of boot between the nags, and heaving a sigh from the very bottom of his heart, "said, bein' as it wus me, and it wus as it wus, he'd take a hundred dollars to boot." Then I felt sad, when I realized this valuable foal was out of my reach, as I had only ten dollars my father had given me to defray my expenses.

I confessed to him I had only this small amount with me. His sympathetic nature seemed to have been touched at my candid statement, as we came to where our roads separated (mine to Jarrett's bridge on the Tugalo river, and his to Pulliam's ferry), he turned his benignant countenance on me and said: "Young man, I see you want my critter, and you ought to have her, I have taken a liking to you, give me the ten dollars and take her. We both lit and changed saddles, shook hands, remounted, and parted to meet no more on this chequered earth. I got to Jarrett's that night, but thought it prudent to get another horse to make the trip to Dahlonega, and on my return mounted the old mare once more. After a most patient ride, I reached home just as the sun was sinking into a molten sea of golden glory, which I construed into a good omen, as it indicated I had made a golden, glorious trade. My father met me at the door, and in some surprise asked me what had become of Francis? I told him I had traded the filly off! I told him in glowing terms of the good luck that had befallen me, of the splendid trade I had made. I expatiated to my astonished parent on the pointed ears, the wide nostrils, the symetrical limbs, the royal blood, the foal, the intelligent color—caught my breath and was about to take a new start—when my father exclaimed "fiddle sticks". I told him that I was not at all surprised at his hastily formed opinion; that more experienced heads than his had been deceived by appearances; told him how the stranger had taken a liking to me, when my parent cried out, "the devil he did." I was about to take a fresh start, when my father shouted out at me "hush;" and not exactly liking the cut of his eye, I hushed. He called up my old friend, black Dan (my former colleague in the pewter buckle mould business) and ordered him to take that old carcass hitched out there, up to old Jake Frederick's, and tell him he sent her to him to have and to keep as a present with a right to all her emoluments, and issue forever; then turned on his heel and left me without another word.

I was greatly shocked at my fathers impatient and reckless manner, but was not set back in my judgment in the least, feeling calmly confident that time, which rights all things, would yet justify me in this horse trade. Yes, I felt as confident of a glorious victory over my parent, as I afterward did in Charleston on the great evening of secession, when I blew my old hunting horn down the streets, that it would be but a breakfast spell to wipe out the yankees.

I made the trip every day up to old man Frederick's. One morning I met the old man at the bars, with a broad grin on his face, and I knew something had happened. My heart fluttered with excitement, as I cried out, all right Uncle Jake? He answered, come and see. I rushed forward with the latin words on my lips, "*veni, vidi, vici,*" and sure enough, there it was: a little, weazelly, mud-colored, sway-backed, crooked-shanked, long-eared m-u-l-e.

I collapsed, telescoped, wilted, and wept for shame. That was the straw that broke the camels back; disgraced, defrauded, heart-broken. This story also got out in the settlement, and to the present day, I have never completely regained my former self confidence in a horse trade.

> " O, some power the gift to gie us,
>   To see ourselves as ithers see us."

MOUNTAIN SPROUTS AND SAND LAPPERS.

The boys from near the South Carolina coast used to call us up-country fellows mountain sprouts, and we in turn called them sand-lappers. Near a little town called Slab town, in Anderson District, off in the woods was a famous school in the days of our story, a large, one-room hewed log house. This school was taught by the deservedly celebrated John Leland Kennedy, who had been a pupil of the famous teacher, Dr. Waddell, and upon whose shoulders the veritable mantel of the Doctor had fallen. Mr. Kennedy was also a preacher, a Presbyterian of the strictest sect, and the word strict would hardly strike those who knew him as striking enough, on account of his striking propensities, for he struck all his pupils in the most striking manner; that he would not hesitate to strike, and indeed, when it was necessary to strike, he invariably struck. His reputation as a teacher was known throughout the country, and he had a very large school, not confined to mountain "sprouts" and "sand-lappers," but boys were sent there from other States, and of all who came to this great school, no boy ever got too big for Mr. Kennedy to strike. Consequently, many unruly boys were sent to this noted school, which increased in numbers so that the house would not hold the pupils; therefore, we were sent by classes out in the grove to study our lessons under the shades of the great oaks. Mr. Kennedy sat in the door, where, with the sweep of his keen black eyes, he could command both the house and the grove, and the least disorder among the groupe would draw from him the sharpest reprimand. Among the members of our class was a sand-lapper from Charleston, named Joe Hide; this boy Joe became greatly attached to one of the mountain sprouts, and would hang upon his words as he related his wonderful hunting yarns, and his hair-breadth escapes amid the wilds of forest life. Joe would listen with wrapt attention and admiration to these narrations, and no matter how extravegantly these stories were manufactured, would swallow down every word as gospel truth. I said the sand-lapper's name was Joe Hide, the other fellow was me, and Joe stuck to me like a leach. I boarded at Dr. Earle's, on one side of the school, and Joe at Dr. Robinson's, on the other side, about three miles apart. One evening Joe decided to go home with me and spend the night; my room-mates were Tom Pickens, from Pendleton, Sloan Benson, from Anderson and John Evans, afterward M. C., from Spartanburg.

Well, after supper, we got up a little game of cards, for fun —we never gambled. Joe didn't want to play, wanted to hear some more hunting stories. I felt a little annoyed at his persistence, when a devlish idea entered my head, and I arose from the table and went to a corner and loaded a pistol with powder, got out an old razor and laid them in a convenient place, returned to the table and at the first opportunity gave the boys the wink, remarking as I did so that I did not feel exactly right; was afraid one of my old spells was coming on me. Joe wanted to know what kind of spells I had. I evasively replied that sometimes I had sorter wild spells, or aberation of the mind came over me, and turning to the boys said, boys if I should have an attack to-night please take care of me, and don't let me do any harm. Joe looked startled and I continued, it distressed me greatly to know that I had done some terrible things while under the power of this awful affliction, but trusted I would not be held responsible for it hereafter, and proposed to go to bed, but Joe was not at all sleepy, and said he believed he would go home; thought he ought to go home anyway. Evans remarked it would be very unsafe to make the trip after night, as two large bears had been seen in the swamp a few days since. Pickens said he would not think of such a thing, as the road was doubtless full of snakes; that in this country the snakes all crawled at night. Joe gave a sort of uneasy grunt, and asked if he could get another room. Benson answered impossible, as all the rooms were occupied, so we all began to undress to go to bed; he got on the edge of my bed, as he was my visitor, and every now and then would ask how I was feeling. The boys kept talking to themselves in an undertone, but every word was audible to Joe Hide.

One of them said what a pity he has these spells, he's such a clever fellow when he's at himself; another one said he's so dangerous, I'm scared. Evans said, warn't that awful about that fellow he killed in Pickens District last year, and they had to choke him off while he was sucking the blood. Benson said the worst thing was his killing that family in Anderson, cutting them up in quarters and salting them down in a hogshead. Pickens said it seemed his whole desire was for blood when those spells came upon him; was about to tell of another terrible affair, when I cried out, boys hold up, I don't want to hear about those terrible things, you know I would not have done it if I could help it, please stop and let's go to sleep. I could hear Joe's heart thumping against his ribs, and he was all over in a shake. I asked what's the matter Joe, got a chill?

He replied no, but he felt mighty bad, and asked how I was feeling now; I told him I felt all right, never felt better in my life, but always before my worst attacks I feel the best. Joe asked if he hadn't better get up and sleep in a chair. I told him no, to go to sleep, that he could tell when the spells were coming on by my jerking. Joe lay still a little while, as if planning for an emergency, suddenly starting up said, suppose I don't wake up when you commence jerking? This last remark tickled me so that to restrain a smothered laugh I made a few jerks before I intended to. Joe made a spring, crying out, " he's jerking boys," and he and all the boys went out through the window; I followed with my pistol and razor, and of course took after Joe. He made for a fence and corn patch hard by, and as he mounted the fence I blazed away with my pistol; Joe and several rails fell on the other side, and as he arose I was close behind him, then through the corn we went, Joe parting the stalks with both hands, as he ran but I pushed him so close that he turned back to the house and as he struck the fence again I split his shirt with my razor from the colar clean out to the end of the tail, making an apron of it; we rose on the fence together and came down on the other side with three or four pannels of fence, but Joe had no idea of surrender, making a break for Dr. Earle's room, bursted through the door, yelling : " Doctor, Doctor, Doctor, he's got a spell on him, he's nearly killed me, oh Lord, Doctor!" The Doctor was greatly startled and his family badly frightened, but he lighted a candle and there stood poor Joe in his wife's room, shaking with terror, his shirt split and entirely open at the back. The Doctor rushed into our room to find out the trouble, but we were back as still as mice, and begged pardon for the disturbance; told the Doctor we had not intended to carry the joke so far, and that Joe got scared worse than we wanted him to. The Doctor laughed and went back to his room after Joe, but could not make him believe it was a joke, nor get him back into our room any more. The Doctor had to set up with Joe the balance of the night and dose him with nerve tonics, and send him home in the morning. Joe never came back to that school again, and I have never had the opportunity of speaking to him since, but I have ever regretted that little cruel escapade, in which I lost a friend and admirer. Joe is still living in that section of the country, where he married and raised a family of excellent children, is doing well and is a solid and useful citizen. He is also better off in this world's goods than I am and no doubt a better man.

HERE'S ANOTHER.

We said Mr. Kennedy was strict, his rules not only applying to our school hours, but also to our conduct at our boarding house. We had positive orders not to go tracing about the country after night, but in violation of this most august authority, we determined to go to a quilting and dance about four miles off in the country. In this spree my room-mates were my companions, the same who had helped me to scare poor Joe. We had lots of fun, and near daylight started home, but when within a half mile of the house, had to climb one of Dr. Earle's staked and ridered fences. We mounted it, and feeling a little tired, sat on top to rest awhile. All at once I said come, boys, let's go, and jumped to the ground, but found I couldn't get up—had broken my leg. The boys picked me up in great consternation, my cousin, Benson, carrying the broken limb with the tenderest care, the least jostle causing me to cry out with pain. They got me to the house with all possible care, laid me on the bed and went for Dr. Earle, who came in, and as I lay there groaning, I managed to give the Doctor the wink (I knew he loved a joke as well as anybody). The Doctor examined my leg and looked very solemn, saying now boys, you see this thing would not have happened if you had obeyed the rules of your teacher; now, here lies this poor fellow ruined for life; this is what is called a compound duplicate fracture of the femur. In plain English, his thigh is broken in three places, and if it aint amputated at once, then

I think he ought to be switchulated. I could not stand it any longer, but sprang out of the bed on both feet in a shout of laughter, when my cousin, Sloan Benson, jerked off his coat and swore he could whip any d—d broken-legged rascal in America, and it took all the boys and the Doctor to hold him off of me until he sorter cooled down; but I soon made it up with the boys, and as we were all in sympathy with eachother about going to the frolic, it was agreed all around to keep the whole matter from Mr. Kennedy.

DISAPPOINTED LOVE.

Love is a powerful thing; it has caused more than one man to leave the roof of his father and mother and go to furrin parts with a conquering female and be exposed to her tender mercy; and agin, it has been the cause of many disappointments, for many fellows have got left, because the cruel female didn't want him to move off with her, for more than many times, she has preferred some other fellow to go along and keep house for her.

The human bosom is also a mighty much of a thing; it is somewhat labyrinthian as to the character of its apartments, and in one respect it is liken unto a reservoir, they both have capacity more or less, the one inginerly holds mud and water and trash and sich; the other holds love and joy, and some other things, not so nice: fear, anger, hate, jealousy etc., and all these latter sins, are called passions, and they go in and come out of a fellow's bosom as they please, when they get the bulge on him. We dont know for certain the number of therooms they stay in; some say the heart, the main office, but we have thought they meander about in the different streets and suburbs of his physicological anatomy, that sometimes they may scoot from the heart to the hollow of the head, and back and forth, and we have seriously considered that they mought at times, show up on the outside of a man, in sich places as the skin of his face, the optics of his eyes, and the very har of his head, selah?

Love is more severer than the grip when it pre-empts a human bosom. It cannot be properly called an epidemic, yet a fellow is likely to take it at most any time before he gets married, and he can take it more than two times also, but the first attack in some cases is considered the worstest, it makes its appearance in some instances, sudden like the measles, then again tardeous, like a cancer, or in a typhodity manner, there is greater danger when it strikes in-nerd, than when it only shows on the outside, in the former case it is much to be dreaded, for it has been khown to cause a man to go plum fool, and has driven some persons to commit death upon themselves or somebody else, and in every case where it strikes the interiro portfolio, that patient ought to be scrutinized carefully, but if it only shows on the superficial areas, there aint no imminent danger, as the fellow will convaless, or get married, which is about the same thing in Dutch.

And once more, there aint so much danger in the passion of love if it visits the human bosom alone by itself, but like some folks we know, it gits into bad company, if it gits in with envy, hate, anger, and the green-eyed monster jealousy, why then there is apt to be trouble, and it is strange they will associate together, fur they never could get along in peace and hominy; I say, when all of these gits in thar together, its like a pack of ravening wolves, and the fellow whats got that bosom is in a bad fix, sure, and there's another thing goes into the human bosom when the door is open, and that is licker, and its a regular ag-ger on of fusses and other devilment, and the more of it that goes in, the worse it is for the fellow and the whole settlement.

It is said that a persen looses his heart when the passions of love gets a mortgage on it, this aint according to our polliticks;

we honestly deem this to be an error, the fellow may lose his gumption, and we solmemnly opine, thats hit; fur how can a fellow lose his heart, when it is hitched on tight to the liver and lights, it seems to me unpossible fur the heart to come out without bringing all the in-nards out with it and that would wind up the whole kerflumux, I imagine.

I remember vividly when the passion of love first occupied my vitals, my heart was unremoved; indeed it swelled up bigger, and a protuberance seemed to have forced up into the thorax, and caused a quite uncomfortable, and choking sensation, but subsequent circumstances have corroborated my convictions that my heart remained in my bosom and continued to perform in a desultary way its accustomed functions, so I still surmise that all this talk about the loosing of the heart is nothing but gass, and also the reports about hearts abursting is bosh and totally onreliable.

But it was not my purpose on this occasion to annihilize the heart, or dissectify the subject of love in all its ramifications, my primoval object was narration. I wanted to norate the history of my experience when I was first conflumicated by this phenomenon, when its phantasmagora first developed in my tender youth.

How fearfully is the human bosom affected when first awakened from its lethargy by the passion of love. I had arrived at the plastic and sweet age of sixteen when its first waves swathed my peaceful breast, when its swashing billows rolled over the component parts of my cupidical system. It was a protracted siege, and for many months helt me suspended over the dark chasms of doubt and hope, until the chords that supported my trembling carcass became frazzled, and at last the weakened strands snapped asunder, and like Lucifer fell flounndering down into the murky abyss of despair; 'twas a stunning fall, and it is still fresh on my memory to-day; that I was deeply impressed at that especial epoch in my life; that something had drapt, and though near a half century has sped o'er the hills and valleys of time since then, I doubt not that some of the footprints of that eventful period mought still be traced amidsts the sands of my gizzard.

I first diagnosed my trouble from the great disquietitude raging in the interior of my internalities, fur indeed, they seemed to be litterally tored up; my old friend slumber became estranged from my habitual command, my grub lost its former savory attractions, instead of my previous eubulistic gush, I was now wont to go it alone, a sort of lonesome far off feeling had sot down on me. I forsook the society of my old hale-fellows well-met, sought the mellowed rays of the sympathetic moon beams. I tried to hold sweet converse with the twinking little stars, fur I felt an aching void in my raging bosom, that nothing else on earth but she could ever fill. But in the vehemence of my desire, my powers of conquest seemed bardaccously parrallalized, and in the persuit of the object of my yearning I became both blind and dumb, and my whole panoply for acquiring victory, want nothing but dern fool ardour. As to tact or diplomacy, I was as guideless as a timble bug, my intantions were noble, but even a superficial observer might have remarked, "his intentions were good, but darn his judgment."

But now I approach the climax, the focus-pokus, the culminating point of the pittiful finale. I had on divers previous occasions made desperate efforts to pop the question, but upon every assay courage had oozed out, and had resulted in signal disaster, when I confronted the fortress. When the time for discreet and skillful action was required that was the very time when my intellectual forces all vanished into the vaguest vagaries, and then too the organs of sound, as well as the muscular attachments of my tongue refused absolutely to correspond. When I was absent from my sweet parralizer, then, my obstinate vocabulary was prolific of words, and my perverse tongue most fluent of speech, and often in romatic groves, and under the somberous shades of giant oaks, I have sat with my finger tips rippling in the limped waters of gurgling brooks rehearseing sweet phrases of my own composition, until I thought I had them engraven upon my memory, and that I was bravely prepared for the next onslaught, but in the presence of those ravishing eyes, all had vanished into a misty dream. To be absent from this adored one, was to me the pangs of death, and my embarassment in her presence, was as a night-horse pressing his horny hoofs upon my smothering bosom, to say simply that I loved her would be an imbecility of expression—with extacy I would have died for her (provided I could feel assured no other fellow would get her); I would cheerfully have laid my hypnotized body down on the cold, cold, ground, with her fairy foot upon my neck absolutely content and happy till the crack of doom. What more can I say?

I had now reached the last stage of desperation, and besought the sympathetic aid of a sweet cousin, and through her friendly manipulation arranged a fishing party especially for this, my contemplated coup. With cruel anxiety I looked forward to that coming day as the most important era of the coming age and fraught with the most momentous results that was to occur on the continent of America. It almost takes my breath after all these years, when my thoughts recur to the torrents of anxiety, the earthquakes of apprehension, and the cyclonic palpatations of the heart, as that day, so pregnant with profound results, approached. But at last it came, the morning of the day came bounding in o'er the cycles of time. I donned my best suit of blue geans and advanced to meet the engagement. I didn't sing any poeans of victory as I marched to the front, for I did not feel assured that my march would be a triumphal one. I often repeated to myself the maxim, "Faint heart never won fair lady," and would bristle up considerably for awhile, but soon found that I was no porcupine and that my bristles wouldn't stand worth a cent. O, how I longed for some such a backbone as is possessed by our ex-President Cleveland, still I proceeded and actually assembled with my Dulcina, and started as her escort to the fishing grounds, but right here my memory fails me. I must have made the intervening space in a walking swoon, for when the first glimpse of consciousness returned I discovered we two standing on the banks of the creek, with a couple of fishing poles tucked under my arm, but I had lost the hooks and the bait; a haloe of glory seemed to surround me, the opportunity had come. I assayed to speak and tried to concentrate all the ideas of my soul into one grand captivating address; that's what I wanted to do and that's what I tried to do, but suddenly the old panic seized me and again I flickered, and the only thing I could think to say to her was, "did you get a bite?" When she naively, but sweetly replied, "How could I without a hook?" Just then I caught another spark from her electrical eyes and went off into a new swoon. I can make no report of what further happened, until the next return of consciousness, when I found us in a pathway on our return to her home; the parental mansion was in sight and my opportunity perhaps lost forever. This fact aroused me to a a sense of the situation, lost, lost, lost, seemed to ring through my burning brain; the extremity of the case impelled me to heroic action, it was the "denier resort," and resolved to plunge the rubicon, suddenly exclaimed aloud, "the die is cast," every muscle of my body taught with excitement, with distorted features, the fingers of both hands run through my disheveled hair, blood-shot eyes, nearly popping from their sockets, I sprang like a wild varmint in the pathway fronting the angelic damsel, and with the yell of an Indian, cried: "Stand, brave Saxon, stand." She gave one glance of horror at my demoniacal expression, then sprang past like a frightened deer and fled to the safety of her home.

I stood transfixed, rooted, grounded to the spot, and would no doubt be there yet, a pillar of salt, but for my affectionate cousin, who sought and led me from the fatal spot.

I never tried it again by word of mouth, but after a lapse of time I wrote her a beautiful epistle, which she answered promptly, saying her hand was pledged to another, but would ever esteem me as one of her most devoted admirers. Then I felt my sandy hair would surely go down in sorrow to an early grave; but it didn't, I'm still on deck, without an apparent scar, married another girl, have lived with her thirty-five years reasonably happy and content. Amen.

SHIRT-TAIL CANYON, CALIFORNIA.

WHILE there is nothing wicked or improper in this narration, yet it is with a feeling of temerity and shamefacedness that I pen it. This sketch is not intended for the eyes of the ladies (for men only); had rather they would not read it, but if through curiosity any of them should do so, then they ought not to call us hard names.

Shirt-tail Canyon is a deep, rough gorge, more than a thousand feet down to the creek bed, and the gold-diggers in descending its rough slopes have to submit to frequent falls and slides, in consequence of which the seats of their corderoys suffer divers abrasions, and in the course of time and the frequency of these slides, the lower or other end of their shirts become exposed, and not unfrequently even obtrude from their proper indoor position, and this common custom of dress from this particular section gave to the canyon this inelegant name.

It was an inconvenient portion of the mining region, the nearest trading point being Yankee Jim's, a smart mining town, and it got its name from the hanging of a man on the spot, of that name. It was no little job for the boys to get in and out of Shirt-tail Canyon, therefore they made as few trips as possible, so when not engaged in work they were often taxed to find amusement; of course gambling of every kind was prevalent throughout the mines, and the most common and convenient games were with cards, but tiring of cards they would sometimes resort to racing, not horse racing, for a horse cannot be got down into that abyss alive, nor could a race track be secured in that rough region; not foot racing, for even a man had to pick his way down there, yet the boys had racing, "kritter racing;" the racers were "body kritters." The boys would select from their persons the best bred of these little fleet footed fellows, put them in a tin plate, push a coal of fire behind them, when around the track they would gallop at an astonishing pace to those unacquainted with the sport, till the third, or any number of rounds agreed upon by the bettors, and of course the foremost "kritter" won the race. Often large bets were made on these pets, and some of them have frequently been known to win their weight in gold dust.

Californians did not regard these little "kritters" with the prejudice and disgust that seems general in the Eastern States. These body "kritters" were simply unavoidable in the California camps; they were indigenius to the soil and climate, and indeed usage and customs have much to do with our likes and dislikes in different parts of the world. The French like horse steak, the Chinaman rat pies, the Dutch dog sausage, the cannibal human flesh, and the American eats the filthy hog. In the California diggins in those days, if one of the boys were asked "what's the matter—fleas?" He would invariably reply, "No, I'm not a dog, its nothing but a louse;" but we will not dwell longer on a subject so repugnant to our present countrymen.

The mining section of California seems to have been an elevated plateau intermediate between the Nevada mountains and the lower plains adjacent to the Pacific Coast, and the heavy winter rains causing the torrent streams in their descent to the plains below to cut out these deep gulleys or gorges, and in the washing down of the veins and pockets the loosened particles of gold are carried down to the creek beds and deposited in the lowest bottoms, though sometimes pockets are only partially interfered with and left on the sides of the slopes, and have often been discovered by the miners far above the present water beds. One of our most expert miners was Jim Kennedy, from Lumpkin County, Ga. He was considered authority on the subject of gold mines, and he had often pointed out a spot to be seen from our camp, where he believed was a rich pocket, yet his faith was not strong enough to make the test, which would require a considerable amount of labor.

One day a raw Dutchman made the descent into Shirt-tail Canyon and wanted a gold mine. We sent him to Kennedy, who directed him to this spot and gave him directions how to work it. The Dutchman went at it faithfully and confidently, worked day after day, and would bring down his pans of dirt to the water to wash for the gold sign. One evening he came down as usual with his pan of dirt, and on washing it out discovered he had made a rich find of gold. Kennedy tried to buy him out or an interest, but he wouldn't sell worth a cent, not even a partnership, and after a few months exhausted the pocket, and went home with more than $20,000. I heard of another Dutch cook who knew nothing whatever about mining, who went out one Sunday morning with a pick and found the biggest piece of gold ever found in the State, sold it to a man for $40, and which proved to be worth over $7,000. I saw the lump or nugget on a gambling table in Sacramento City, about a year after the find.

CHASED BY WOLVES IN CALIFORNIA.

Becoming tired of the gold mines, and desiring a change, I determined to go to San Francisco, and shoot wild fowls for the city market. On the boat from Sacramento down the river, I heard some passengers speaking of Noble's ranche. They thought he was a South Carolinian, and I was satisfied it was my old friend and schooll-mate, Alick, a son of ex-Gov. Noble, of S. C. Many a time had we hunted together in the mountains of N. C., so concluded to change my purpose for the present, and pay him a visit. From San Francisco I took a small steamer to Petaluma, and from thence not being able to secure a conveyance, after receiving instructions as to the route, struck out on foot, a distance of sixteen miles.

I always carried my shot gun, and my way leading some distance down a creek, discovered a number of fine English ducks, and soon became engaged in shooting and packing my game, oblivious to latitude, or longitude, or the flight of time, nor did the waning hours occur to me till late in the afternoon; nor had I paid any attention to the instructions given me as to destination, was satisfied I had wandered considerably out of my course, so gave up the sport and turned my attention to the seeking of my friend's home, still hoping to reach it before night fall; but dark overtook me trudging along, with a heavy pack of ducks swung across my shoulders, tired and foot-sore. I was thinking of my far-off home, of friends and comforts left behind. A sudden feeling of desolation occupied my mind, and a premonition of some coming evil depressed my spirits. My reveries were broken by the sound of a quick yelp, and a succession of weird howls wafted on the still night air. The truth flashed upon me as an electric shock, wolves! I had heard the sound before. I knew the gray wolf of the Blue Ridge mountains. I had read of the ruthless red Russian wolf, and of human bones bleaching on the plains. I was at first startled, now alarmed, for I knew they would soon be upon my track; all my fatigue vanished, forgot my skinned heels, and quickened my pace into a trot. Directly the yelps multiplied and constantly became more distinct; they were on my track and getting nearer every minute; were rapidly gaining on me; that my pack of ducks were impedeing my speed, I had not thought of before. I dashed the game on the ground, and spurred into a gallop; but on came the wolves. I could hear them scrambling and fighting over my game, but I did not stop to listen—but got faster—my whole mind and body concentrated in one grand central idea: to get away from those wolves if there was any possible chance to do so. I revolted at the thought that my poor carcass should make their next repast, as I sped along the plain. I saw off to my right a huge rock rising in the prairie. I turned my course thither, and as I drew nearer found its front inaccessible, but running round discovered a water rut, and up this groove I climbed to the verry top and stopped, because I could'nt get any higher. By this time, the terrible brutes had finished the ducks, and were again on my track; nor did I doubt for a moment that they wanted the man who killed the ducks. I at once began preparations for defense. My gun being empty, I quickly threw down the old muzzle loader a couple of charges of powder, rammed down the wads, seized my shot pouch, when to my horror, discovered that

I had galloped all the shot out of my pouch, and now the wolves were around the rock, and had set up the most hideous and hungry howl. I could see their dusky forms and gleaming eyes—they seemed unusually large—larger than the gray wolf of the Blue Ridge, and perhaps more ferocious than the terrible Russian wolf. Now, two of the daring devils start up the very rut that I came, and as they get in a few feet of me, I let both barrels of powder off right in their faces. This was a surprise, they rolled down the rock and the whole pack scampered away. Just then a flash of light caught my eye out on the prairie—some body's home. I slid down the rock and made for the light. I never felt so agile before in all my life, as I seemed to fly over the grassy land, and during that brief transit, the panorama of my whole life passed before me. My thoughts traveled from childhood to the present and on into the next world to the judgment to come; but all this did not impede my speed in the least. I believe that extraordinary physical action produces extraordinay mental action, and *vice versa* It is retroactive, and just then I felt assured that all my mental and physical forces were pulling together. I knew I must annihilate time and space, or be annihilated myself, for the ravenous beasts were again on my track. They were coming, closing up the little space between us. I could almost feel their hot breath and bloody fangs, rending my limbs and gnawing into my vitals.

I reached the house first, the bloody wolves at my heels. I did not try to ring the bell, and gave but one knock at the door; knocked the door clean through, both falling inside, but I was on top of the door and had burst into the kitchen. The hurricane awoke the cook, a Sweede, and he was scared nearly as bad as I was. He could not speak a word of English, nor could I speak a word for want of breath; for sometime we stared at each other; but, oh! I felt so thankful that I was safe, when sufficiently recovered. I tried to make the Sweede understand my narrow escape from the wolves, and just about the time I thought he was catching on to my signs, the fool commenced laughing as if his very sides would split open, and then I feared he had lost his reason from fright, but after a spell he checked up and I succeeded in making him understand that I was Noble's friend: he indicated to me that Noble had gone out but would soon return, and gave me a lunch and showed me a cot. I could not sleep after so much excitement, and my heart was so full of gratitude that my unworthy life had been spared by what appeared to be a special interposition of Providence. As I lay there in the cot, I reflected much and tried to repent of my many misdeeds and shortcomings, and formed many good resolutions for future conduct.

Late in the night I heard my friend Noble return, heard him talking to the cook and heard them both laughing; after awhile Noble came in and recognized me with great joy, and as soon as our first greeting was over I tried to tell him of my wonderful escape from the terrible wolves, when to my astonishment and mortification he broke out into a hysterical fit of laughter, and laughed until the tears rolled down his cheeks. He saw that I felt hurt and as soon as he could recover his faculties he said: "Why Dave, they were nothing but harmless coyotes, and were never known to attack a man." I told him how large they were and how their eyes gleamed on me. He said: "No, they were not half as large as the Blue Ridge wolf, and absolutely harmless."

But those pretended wolves had hurt my feelings and I determined to be avenged upon them, and before I left the California prairies, killed as many of them as I thought tried to kill me.

A RABUN COUNTY, GA., FROLIC.

In the olden times, dancing was by odds the favorite amusement with the young people, and in my youthful days I engaged in all kinds of terpsichorean felicities, participated in the fashionable cotillions, waltzes and polkas, at the balls, weddings and parties, with the elite of that day; have been to the piney woods frolics, shin digs and stag dances, but in Rabun County, Ga., where once lived our Chief Justice Bleckley and the silver-tongued H. V. M. Miller, I attended a frolic, that for intensity of enjoyment, cast a glamour over all the balance of my experience.

I had recently returned from California, and my father was a contractor on the old Blue Ridge Railroad, in South Carolina, and had taken a contract in Rabun County, Ga., known as the Whitmire fill, and said by Col. Walter Gwinn, chief engineer, to be the deepest railroad fill then known, measuring 108 feet from the culvert to the top of grade, and a description of which was given by our Judge George Hillyer in an Athens paper, in his youthful reportorial work, and where I first made his acquaintance.

In this contract I was to be a partner as well as a manager and had made a horse-back trip up into Rabun. I was riding through the rich valley, at the very head-waters of the Tennessee river, with a resident young man named Major Gibson. Late in the afternoon (dusk had already commenced to throw its sable mantle over the beautiful valley), as we passed a store we were informed of a log-rolling and quilting close by, and decided to attend; but as we had not participated in the labor of rolling logs, and did not like to intrude without some equivolent on our part as a contribution, so bought a jug of mountain dew and had it sent over to the frolic; we were welcomed and our present was well received by the boys. We were introduced as the men from Californy, and we all took a familiar smile from the afore-mentioned jug.

The quilts having been finished and removed, the frolic had already commenced. Our host, Jack Bradley, was the fiddler; his favorite tune was an old-time famous one, and widely known as "Rye Straw," and Jack's performance was entirely confined to the bottom part of the tune, but after a bit (like the Arkansaw Traveler) I ventured to ask him if he never went up stairs on that tune? He answered he didn't, because he didn't know where the steps was, and handing me the instrument asked if I could play the fiddle? I answered that sometimes I sawed a little and put the upper story on "Rye Straw" the best I could. It proved a ten strike, as I soon discovered that I had become a very popular person. I showed Bradley the stairsteps and soon had him educated so he could go through the upper story of the tune.

Suddenly I felt a slap on the shoulder and turning, discovered my assaulter to be a splendid specimen of fresh mountain girlhood, a beauty with rosy cheeks and sparkling eyes. She said, "Californy, less you and me take a turn." 'Nough said, says I, as quick as a cat could wink its eye, and calling on Bradley to give us the best he had in the shop, he promised to empty out the gourd for us, and added : "Go it, Californy, if you keep up with that gal there aint nothing in this valley too good for you." Now, the floor of the house, like many others in that section, was made of puncheons, split out from the forest trees and laid on chestnut or wild locust sleepers, and, consequently, quite springy. Chairs in Rabun County were not then so plentiful as they now are in Atlanta, and it was not an unpopular custom for two of the young people to sit on the same chair together and in a dance, frequently a couple would occupy the floor, especially in a break-down.

I had been challenged by the belle of the valley to single combat and knew I was in for it, but had fully determined to be on hand when she got through. I led Miss Mary D. a few turns up and down the hall, stopped in the center where we made our bows, forwarded and back, swung corners and circled all, crossed over and back, then the fun commenced. I made a pass and she coquetted, I cornered and she chassed, I shuffled and she sidewized, I pigeon-winged and she wire-toed, I double-shuffled and she gave the toe-whiz, I gave a jim-crow lick and she kill-krankled, I struck a break-down and she hit the hurricane, I went into a jig and she jiggareed, and for every lead I'd make she'd call me and go one better; now and then we'd change sides and cross back into another break-down, and it was go it Miss Mary, hurry Californy, and Jack Bradley seemed to have got inspiration on "Rye Straw." Major Gibson beat the fiddle strings with straws, one fellow beat a triangle, several were patting and every gal was keeping time on the floor with her feet, and the heads all around the room were bobbing up and down with the spring of the elastic floor. Now and then some chap would sing out, "go it frolic, yer dady's rich and no poor kin;" "hurry Miss Mary, come down to it Californy," and we were both doing our very level best. Miss Mary was a picture—to say she looked a thing of life would be but a feeble and emaciated expression. I can still see her after the lapse of time in the midst of one of those dead setto's, her lythe and willowy form swaying from side to side in a quiver of action, athletic and graceful in her very motion, head and shoulders a little inclined to the front, the folds of her blue-checked homespun frock grasped in her hands on either side, a little raised to clear her shapely ankles, her skirts artistically spread out and in, to a perfect harmony of motion, and her dainty feet would strike that puncheon floor with the quick beat of a knitting-machine, and she skimmed the floor as smoothly as a full-rigged brig, as she cuts the great deep, rocking from side to side before a spanking breeze (talk about your germans of this advanced day and of the enjoyments of your young folks, all tame to that). And I was right about there too, head and shoulders thrown back to the break-down, a little to the front in the pigeon-wing, arms flying to help the feet keep time to the music; the weather was getting equatorial, the perspiration streaming, and we were just getting down properly to our knitting in what is called the cyclone movement, when the music suddenly ceased. Jack Bradley had sawed his treble string clean in two, and it was a draw between me and the belle of the Tennessee Valley. We retired to a chair amid the plaudits of the crowd; were pretty well blowed and a little fatigued, but I found a delightful repose for my arms, and my partner rested one of hers on her lap and the other round my shoulder. Miss Mary felt a little warm, but not at all disagreeably so; our temperatures ranged about the same degree farenheit. The caloris gradually cooled down to its normal state and we spent several very agreeable moments together watching the other couples as they would take a turn.

"We danced all night till broad daylight and went home

with the girls in the morning," and as we passed the store treated the girls to torter shell side combs and sacrament wine.

The Miss Mary D., of the valley, is still there, but now a silverhaired matron and the faithful mother of a crowd of excellent children. My locks too have changed to a frosty hue, though now and then I still saw on my old fiddle and never strike old "Rye Straw" but I think of Miss Mary and the Rabun County frolic, and when I compare the good old usages of those days with the present fashionable arm clutch, it is impossible for me to restrain a feeling of contempt.

MRS. JULIA A. SLOAN.

THE VICTIM.

I HAD profited by experience: my second effort was adroitly and ably managed—was an eminent and triumphant success in every particular. I had now reached the more mature age of twenty-six years, and had become more rational in many respects; and while it still gives me pleasure to crow over this victory, I want to tote fair, and right here make the acknowlegment that my victim was captured at the tender and inexperienced age of fifteen, and of course more susceptible than one more advanced in maidenhood. This avowment, I am aware, will detract a part of the glory, but I have still enough left to make me feel comfortable, and, besides, after the lapse of thirty-five brief years, have got the old girl to boot.

I had long ago become reconciled to my first disaster, had now and then sparred a little among the girls, but nothing serious had occurred. Now I began to realize the need of a help-mete; wanted a good wife, craved the refining influences of a good woman, to pare away the rougher protuberances of my nature. I caught the idea from the poet Tupper, and want him to have the credit of it—to pray for a wife, to pray for a good wife, just such a wife as the Good Lord knew I needed. I did pray in faith, believing, and watched for; and to-be-sure my prayer was answered in such an especial manner, to my mind, as to leave no doubt that it was in response to my earnest supplications. I met with my fate at Anderson, S. C., in the person of a college girl, who was so pointedly flung in my pathway that I recognized in her, at once, the providential boon. As pretty as a pink, as gentle as a dove, and sweeter to me than taffy. I gathered all my forces and stormed the citadel without delay, allowing no grass to grow under my diligent feet. I went in to win, and the recollection of my former weakness only made me the stronger and more determined. We first commenced playing the fiddle and piano together, and accorded from the start. Our music made others merry, and we had a little side-show of happiness to ourselves; and although now thirty-five years have sped since our union, I keep up my fiddling with an harmonious accompaniment from this dear old girl, and, if spared, hope to keep attuned to our golden wedding.

I felt from the first, in this campaign, that I was master of the situation, and talked out my love like a little man; and my love listened to me lingeringly, and like the fellow who had no heart to refuse a drink, gave me her hand affectionately, and referred me to the old folks. But didn't I feel good then. How I stepped around the streets of Anderson next day. I stood so straight in my boots that I sorter leant back. The girl was mine, and I didn't care who knew it. I thought of my first failure and how greatly I had improved upon that effort, how nice I had done up this job, and that I was no longer to be classed with the batchelor dogs; soon to be a respectable married man, the prospective head of a family, a man of responsibility and consequence, and no telling what the future might have in store for me.

The greatest trouble I now had to contend with was impatience. I didn't want to wait. The little lady lacked about two years of finishing her education. But I decided it would be better for the education to be disappointed than me. I wanted the girl, and wanted her quick—like Judge Bleckley wanted the earthquake to stop.

Soon after her vacation, I followed her to her home on the Saluda, in Edgefield; got off the railroad at Chappell's, and, fortunately, met a gentleman taking the train, who kindly tendered his large coach and baggage cart to take me to my destination, as it was near the route home. Some time after dark had set in, we crossed a bridge and drove up a long rocky lane approaching the mansion. It was one of those close, sultry summer evenings, so common in our southern climate, and the rattling of the cumbersome wheels of our vehicles seemed to make the most extraordinary lumbering noise. As we drew near the dwelling I observed, in the lighted windows, numerous heads poking out, to ascertain the cause of the unusual rumpus. Our caravan halted before the front gate, and your writer descended and walked with a stately tread to the entrance, and was met there by a brother of my intended. I introduced myself and announced the object of my visit, and, upon invitation, resolutely moved forward into the parlor, filled with people. A single glance satisfied me that I had interrupted their evening devotions. A reverential old gentleman was peering over his spectacles, with Bible in hand, and beside him sat another old gentleman, who I decided to be my future father-in-law, and the balance of the company to be guests and members of the family. On the right, I discovered my jewel, greeted her warmly, then moved around the room with my usher, bowing in the most gracious manner as introduced, then modestly retired to a seat in the rear of one corner of the piano and listened devoutly to the family service, scarcely casting a glance in the direction where my eyes especially desired to range. The service over, my attention was directed to two young ladies, who were talking in an undertone and casting significant glances in my direction; then heard a suppressed giggle as a servant girl placed a lamp on the corner of the piano so as to shine directly in my face. Turning suddenly to where my sweetheart sat, the seat was vacant; she had slipped out. Then my old experience recurred in all its original force—deserted! She had gone back on me: and now, like Peter, I lost my faith—all my prayers for a wife wasted. Right then it occurred to me that I or the Lord one had made a bad mistake; and to complete my disgrace, I was now an object of sport for those two young ladies. It was too much. I determined to forget the unfaithful girl and my prayers, and to leave for home the next morning—even thought of hunting another roof for the night; but the girls had riled me. I was going to get even with them before I left, to show them I wasn't the kind of a bug-eater they took me for; so I picked up my chair and sat down right in front of them and commenced a rattling, don't-care sort of confabulation. This bold act brought them to their p's and q's, and placed them on the defense. Directly, turning around, I saw the two old gentlemen regarding me, as I thought, with critic's eyes, and feeling just like I didn't care what corn was worth a bushel, moved my seat and tackled them, and soon had the pleasure of thinking I had paralyzed the whole party, and was now ready to vamose the ranche, cursing (in my mind) the unreliability of the entire fair sex.

Just about this condition of affairs I observed a handsome young woman tripping in the parlor; the gay deceiver was making for me. I recognized the face and the form; was she about to tell me to git up and git? I braced myself for the coming shock, and it came, but not as I expected, for she sweetly invited me in to take a lunch. I had forgotten that I had no supper. She sat by me as I partook of her hospitalities (both she and her mother), and as I was satisfying the inner man I also saw through the millstone and wilted, and as my ire abated so did my glibness of speech. It got real hard hard for me to think of anything at all to say, and once more recurred to me my old time predicament. I got too modest to talk of love that night and retired from the scene under considerable embarrassment, but got all right next morning when I learned that my faithful one had reserved a buggy for our especial use to ride to church. My tongue soon recovered its wonted roll and I remained pleasantly situated for several days. In fact, I felt loth to leave at all, but my embassy was unfinished; I wanted to secure the consent of the old folks. At the first opportunity I invited the old gentleman to take a walk, and when well out the gate he asked if I would like to see the crops. I answered abstractedly in the negative, then he proposed the meadows and the stock. I answered, "No, sir, not on this occasion, I am here on an entirely different business, and doubt not, sir, you have already guessed the object of my mission. He answered, "that it was not his custom to guess at other peoples' business." I must have looked surprised, but frankly told him what I had come for, and how gladly I would receive his consent to my suit. He remarked that I was a stranger to him and that he would like to be better informed before he could give his daughter to a stranger. This stumped me, but only for a moment. I proposed to the old gentleman to go home with me and investigate, but he made the objection that he could not leave his crops for the present; then I asked him which was of the most importance to him, his present crops or the happiness of his youngest daughter; this got him, and he went. Now I had to talk to the old lady, and managed to find her in the parlor early next morning, so brought the question to bear at once; but she said she wanted her daughter to learn more about the responsibilities of housekeeping before she married. I told her I had an old mother where I lived, who was the best in the world about that. She then remarked very positively that her daughter was too young to marry; but I was posted and asked the old lady how old she was when she married, and this was a sockdolager. Then, as a concluding remark, she was not willing to give her daughter up; but I spiritedly told her that was exactly my fix, that I'd die before I'd give her up; then the old lady fled, and I never did get her consent.

But I took the old gentleman home with me, and he and my parents consulted together over the matter, and after the caucus had been held I was summoned to the parlor. My father was the speaker; he stated that the matter had been fully discussed between them and had been decided in my favor, but that they had all agreed that it was best to postpone the marriage for two years longer and allow the young lady an opportunity to finish her education; then I kicked, kicked the tea overboard. I took exactly the position the confederacy took toward the federal government, secceded, positively refused to accept the amendment, called for the previous question, and, like Tom Reid, counted the votes all my own way. I firmly stated to my seniors that the time was irrevocably "sot," and we were of the opinion that as we were the parties mostly interested that our decision was a matter of much consideration, and we carried out our programme. The dear old folks against whom I rebelled have long since gone to the better world, and we are whacking along here yet through the rough lanes of life, and have had many ups and downs, but mostly downs, and have long since tried to learn submission to Him who hath joined us together; and though our locks are growing whiter each day, we still feel young and fresh in heart; and as we approach the shadowy end, are filled with the hope and trust that though we shall soon leave this earthly scene of many grievous trials, that after a brief separation we shall meet again and continue our journey together through the spacious halls of eternity,

> "There to bathe our weary souls
> In seas of heavenly rest."

FALLING OFF A MOUNTAIN.

IN front, and in view of our old mountain home in Fairfield Valley, N. C., stands the Rock Mountain, with its bare walls rounding up a thousand feet towards the sky; on its summit is an extensive area of ravines and ridges, covered with the native forest tree, and used to be a favorite tramping ground for the deer, and I have killed a number of them started on this mountain. On one occasion I went up on this mountain to hunt alone, except my dogs, and soon a deer was sprung, but loth to leave the mountain, it played around ahead of the dogs. I was slipping along trying to get a shot when I saw it coming clipping along toward me where I had stopped not far from the brink of a precipice. The doe discovered me too late and attempted to pass between me and the precipice, when I fired and gave it a mortal wound. To my astonishment the wounded deer turned abruptly and went headlong down into the abyss below. I rushed forward to the brink to peer over and see where the poor thing had gone down, when to my horror my heel slipped and over I went after the deer. I remember closing my eyes, for I knew it was all over with me, and I also remember as I started, my first thought was of prayer and that I would have to make quick work of it too. I think I had got about as far as "Now I lay me down to sleep," when I brought up with a sudden jerk and thought I had struck the bottom and was a dead man, but in a moment reason began to return and it occurred to my mind that I had made the trip too quick. Then the question arose, was I dead or not? whereupon, I opened my eyes and discovered that I was alive and unhurt, sitting astride a clump of ivy bushes that grew in the crevice of the rock. My hat had gone over after the deer, but I was sitting safe enough astride of those blessed bushes with my gun still clutched in my hand, and looking up discovered that I had slid down about fifteen feet from the top of the rock; but how to get back I did not see. I could finish the trip down with very little difficulty, but was not willing to make the trip voluntarily, to the contrary hugged the rock at my back more tenaciously; indeed there has ever been an inclination in my nature to ascend rather than descend, though in actual experience I believe the latter has been my fate.

But the all important question with me now was, how to get out of that place. I was discontented, was dissatisfied with my position in life; I wanted to resign and even to abandon the position without a formal resignation. Oh, how I needed the advice and aid of some good friend just then. From my sit-point I could see into the veranda of my own sweet home (had been married but a short time). When in great trouble I try to reason as well as pray—and reason as methodically as possible. To extricate myself from this terrible imprisonment, I had to devise some method, so I adopted methodism unanimously, and began to shout most lustily; but it soon occurred to me that there was a difference between the Methodist's experience and mine, for they claim to shout when they are harpy, and I felt sure I was not happy; I did not feel the slightest symptoms of happiness, still I kept on shouting, but it was no go, for the wind was against me and I could not make myself heard. I continued to shout; to tell the truth, I yelled and yelled until my voice broke up in utter hoarseness.

I saw my young wife come out on the veranda and look towards the mountains, as if expecting to hear from or see me; and, oh! how I longed to be there. Home never looked sweeter to a living man than mine did to me then. I thought of the good old song, "Sweet Home," and tried to sing it, but had got too hoarse to sing. In fact, I did not feel much like singing anyhow. After awhile, I saw my darling turn and go back in the house; then a feeling akin to that of Mr. Selkirk's took possession of my poor, isolated soul. I wanted to go home. I wanted to be more social; wanted to be an affectionate husband, a good democrat, an exemplary Christian, and get something good to eat; but the unpleasant fact stared me in the face that I must get out of my present predicament before I could do or get anything. My wife came out again and looked anxiously, and I returned the look with double compound interest ; but, alas! she retired again. I remained in this awful position three weeks, thirty-seven days, forty-two hours, sixty-five minutes and ninety seconds (at least so it seemed to me). At last a negro man named Jim came into the cove below to get white oak splits, and I succeeded in making my position known to him. I directed him to come around to the top of the mountain above me and cut a long pole, with which he pulled me up to the point from whence I started, and was thus delivered from my perilous position. My deliverer was Jim Hacket, one of our slaves, and I have never seen the day, from that time till now, that I would not cut my tobacco right in the middle and give the biggest half to that old darkey.

THE ANXIOUS ENQUIRER.

From the surrender at Appomatox, I returned to Edgefield, S. C., where my wife had remained during the war, with her father. I had sold all the property I possessed, except negroes, at the beginning of the war, and invested in Confederate bonds; now, the war ended, I found the bonds worthless and the negroes free. I had three silver watches, $7.50 in silver and $15 in greenbacks, captured from the enemy when I was a scout, and my horse. I swapped all but the cash for a wagon and team from Johnson's returning soldiers, and moved with my little family to Southwest Georgia, to start anew.

I bought a plantation, with outfit complete—stock and implements—on a credit, from John W. Jordan, Jr., near Smithville, in Lee county, and started the business of cotton planting. I had been raised, as we then thought, above the cotton-belt; although my father used to plant several patches of cotton, I knew nothing of its culture. I had sold out a splendid stock farm, to go to the war, and my teaching had been to raise grass and not to destroy it.

In starting a new business, I thought the best way to get at it was to obtain all the information possible of the *modus operandi* of planting cotton, and so set myself to work visiting and pumping my neighbors for the coveted information. I made frequent visits to the Wellses, Jordans, Jenningses, Jays, Allens, Birds, Rosses, etc., and as Lem Jay (who was considered a crack cotton planter) remarked, got everybody's opinion then turned around and done as I durned pleased. There was a scarcity of cotton-seed in the country, and difficult to secure even at a high price, and it had become to me a question of great perplexity.

One Sunday, I went with my family to spend the day with Mr. William Wells, and found there quite a number of neighboring planters. We were all sitting out on the front veranda, and, as usual, I was spunging out of the party all the information I could get, when that scamp I referred to before, Lem Jay (and who had seen me the day before setting out cabbage plants) remarked that he thought he could put me on to a plan that would interest me. He said that Mr. Jule Bird, a neighbor and very large planter, had a great deal of cotton already up, that it had come up very thick, and he would commence chopping it out to-morrow, and had no doubt that Mr. Bird would take great pleasure in furnishing me all the plants I might want, free of charge. Mr. Bird was present and said it would afford him great pleasure to do so, and that I would be welcome to all I wanted. I expressed unbounded gratitude to the gentlemen for their kindness, and was about to make a little speech of thanks when I caught a glimpse of several roguish looking winks sliding around, and stopped suddenly short, as I smelt the fumes of a dead rat, when there followed a general explosion of risibles at my expense; but full amends were made by their assistance to procure the necessary cotton-seed. One day I called on my neighbor John W. Jordan, Sr., and had plied him with many questions on the cotton making business, and finally asked him how many bales he thought I ought to make this year. He surveyed me solemnly from head to foot replying that he could not tell me how many I ought to make, but if I made airy bale he would be mightily surprised. The shock was a severe one to me, but I had the pleasure of beating the old gentleman that very year; and I think he was sorry for the joke afterward, when he found out that I was the son of his first sweetheart, whom he had very earnestly courted in his younger days.

I was greatly puzzled when my cotton commenced blooming to find both white and red blossoms on the same stalk at the same time; why one should be red and the other white, I could not get at the philosophy of it; the chemical action on the part of nature I could not quite understand. This brought another good laugh from my neighbors, and the discovery to myself that I had gone off in this instance half cocked, for had I waited and observed, would have learned that the blossom is white the first day and red the next.

But here's another rigid joke. One day I was sitting on the fence watching my hands hoe cotton, when a stranger to me drove up and alighting from his buggy took a seat beside me and commenced conversation (the whole thing was a put up job). After awhile he said he had been driving around through every part of the country and had never, in all his life, seen such grassy crops (it had been a very rainy season); but he'd bederned if I wasn't considerably worse off than anybody he had yet seen. This hit me heavy, for I thought I was ruined, and as soon as the man left I got a hoe and let in and whooped up the darkies and got rid of the grass, but the unaccustomed exertion cost me a spell of fever. I made one of the best crops to my force in Lee County, that year, and fully established myself as a cotton-planter.

Now for the benefit of despairing humanity I will tell an anecdote on a Lee County young man, who, at the time of which we have been talking, was my neighbor farmer; our places joined and when I left that country, left him there. Some years after, I came down to Atlanta from Norcross and met my former neighbor on the street. He informed me that he had just arrived in the city and had come here to practice law. I was astonished and asked him what he knew about law. He said he had busted farming and had taken to law; had been studying it for a few months, and asked my opinion as to what I thought of his chances in Atlanta. I gave him my opinion candidly and in a flat-footed manner. I told him these Atlanta lawyers were a sharp set, and the chances for a country fellow who had come to a great city with a smattering of law was about as slim as anything I had ever seen. His face lengthened out as I talked to him, and finally he exclaimed: "I am obliged to succeed; I've got nothing but a family, and it's a "ground hog case," and stamping his foot in a resolute manner, said: "I am obliged to succeed." Then I said, Bob, if it has come to that, go ahead and maybe you will; "where there's a will there's a way;" and if you are obliged to do it, you will. That same fellow is familliarly known to almost everyone in the city to-day as Bob Jourdan, and one of its most popular lawyers.

HOW I GOT RID OF PRICE ALBERT.

NEAR the famous Cashier's Valley, in the Blue Ridge mountains of North Carolina, and two miles across a gap, nestles as lovely a little spot as this noted range can show, Fairfild Valley, resembling a great ampitheater, with its lofty blue rock walls surrounding.

Here my father's family used to spend their summers, and here I afterwards, with my uncle, J. T. Hackett, ran a stock farm and a summer hotel. We raised cattle, hogs, sheep and mules. Among other animals, we owned an imported jack named Prince Albert, that cost eight hundred dollars. After a while the confederate war came on, and we had to abandon this lovely home, and went as volenteers to fight our country's battles. We sold out every thing except this especial animule. Not being able to find a purchaser for his royal highness, I sent him down to Edgfield, S. C., and boarded him out during the war, and when the war was ended moved to Southwest Georgia. Still not being able to dispose of the prince, I transported him, at considerable expense, to my new home, where he became not only a considerable expense, but a nuisance to the whole neighborhood. He would not bear imprisonment, either by fences, bars or gates. Not satisfied with injury to my own property, he committed depredations on my neighbors. The more I tried to sell him, the more I couldn't do it. Finally I tried to give him away; couldn't even do that, and indeed this jackass problem had become one of great anxiety and gloom to me.

One day I had my hands near the public road, raising some timbers to build a carriage house, when I heard a halloa out at the road. I turned and saw a solitary horseman halted in the highway. He called to me in the most beseeching tones, and said: "My friend, will you be so kind as to step this way, just for a moment." He seemed in great distress, so I ordered the boys to stop work till I returned. As I approached, the man reached out his hand and grasped mine, saying: "My friend, I want to ask a favor of you: do not deny me; I am suffering." I asked, "What can I do for you, sir?" (feeling my heart melting toward the poor fellow.) He continued: "My good friend I have been riding alone for hours down this lonesome old 'Bond's Trail.' I have not met or seen a human face, and I am under a most sacred vow. I have sworn never to take a drink of spirits by myself, and I have in my saddle-bags some of the best old peach brandy you ever wet your lips with. I want you to take a drink with me; please don't refuse, for I feel I cannot stand it any longer." The favor seemed so small and the self-denial on my part so insignificant, that I complied with his request. Then he took the bottle, and a goodly portion of its contents went down his thirsty throat. I then offered him my hand and wished him a pleasant journey on his way; but he held my hand, and pleadingly said: "My dear friend, don't go yet; just one more, please." I took the flask and turned it up to my lips, as if I intended to take another, and—did, then, after watching him gurgle down swallow after swallow, begged to be excused, as my hands were waiting for me, and again bid him Godspeed on his way, when he cried out: "Oh, my dear friend, my good friend, just one more before we part. His tone was one of abject entreaty, and to get rid of the man, I smiled once more with him, and said, "good-bye, good-bye, sir." As I walked off, he watched me regretfully, hailed me again, and said: " I, say, my friend, have you got anything to trade?" I stopped, as my troublesome mule flashed across my mind, and answered, "Yes, sir, I have a very fine jack that I would like to trade." He said, " bring him out," at the same time drawing another flask; then handing me a watch and chain, he added, "I will give you this for him." I did not take time to examine the trinkets, but called to a boy to bring out Prince Albert. The trade was confirmed without further talk, he only requesting that I let the boy go a mile or so, to get the prince well started. I ordered the boy to go with him twain. Before starting, however, he took my hand, and said: "My friend, my benefactor, as long as my life lasts, I shall feel grateful for your kindness to a dying stranger. I was athirst and you helped me to drink. I will never forget you; I shall cherish your memory as that of a friend in my time of need, and now, in this parting moment, perhaps forever, favor me just once more." I favored him, and thus we parted. I watched him and Prince Albert go down that long lane until they passed beyond my sight. I have never heard of either of them since, but have often hoped that both were doing well.

I found both watch and chain to be good gold, and traded them for a fine horse; and since that jackass trade, I have concluded that there is always some hope, even under the most adverse circumstances and the gloomiest out-look.

THE PROPHETIC SPEECH.

ADDRESS DELIVERED BY D. U. SLOAN, BEFORE THE EARLY COUNTY, GA., AGRICULTURAL SOCIETY, JULY, 1874.

Mr. President, Ladies and Gentlemen—This large and respectable audience is encouraging, and is proof that "there is life in the old land yet." The presence of so many ladies inspires fresh hopes for this society. Mr. President, I discern that your benign face wears a more congenial glow, your eyes scintillate with gleams of returning youth, as from your elevated perch you gaze admiringly upon the fair forms that surround you. Our Secretary, too, appears more sprightly, while from his humbler position he steals the furtive glance, delighted that his earnest effort to induce their presence has been crowned with success.

Brethren of the plow, do we not all feel happier for the presence of these fair friends in or midst? May they continue to come and cheer us with their smiles of approval, and help us to promote the great cause of agriculture. And, ladies, please pardon me if I remind you of your great rsponsibility in life, for in all the annals of history, from the first fair maid of Edenville, through the tedious chronicles of generations, till you come to the mother, the better-half, or the absorbing sweetheart of the present day, and behold your potent influence over the so-called lords of creation, for weal or woe; and while the unhappy experience, the lamentable difficulty, of the first sweet girl, in the primitive garden about the fruit, may serve as a gentle reminder, yet remember your influence for good or evil is not abated one jot or tittle. Ladies, ever encourage the worthy enterprises of your infatuated admirers; frown down by your absence all their evil works, and so shall you truly become the good angels of deliverance to your less refined and more obdurate companions of earth.

Mr. President, I have neither the disposition nor the information to discuss the science of agriculture, and will leave such work for wiser heads than mine. I only propose to offer a few general ideas on subjects of vital importance to the class of men who earn their bread "by the sweat of the face," and as the great Mr. Greeley *should* have said, I want to tell you what little I know about farming.

Mr. President and brother "crappers," as sure as I stand before you to-day, without hesitation or reservation, without fear of successful contradiction, and in all the solemnity of truth, I feel constrained to state that the noble, the wonderful, the glorious profession of agriculture has nearly "busted" your humble orator, "enduring the last few craps." But, sir, I believe—and I find much comfort in the thought—that all hope is not yet with me fled, for I believe the right kind of farming can be made profitable. Sir, your eloquent speaker of last month told us how he had come to grief agriculturally. He gave us a graphic description of the romantic castles he had constructed in the azure skies of paper calculation; told us of his fond devotion to that gay coquet, Miss Delila Cotton, and how he had been blinded by her charms and had yielded to her fascinations; how he had tripped the fantastic toe with her in the mazy dances of fortune, whirled with her in the dizzy waltzes of speculation, gyrated in the polkas and highland flings, cut pigeon-wings, and went through all the fancy steps of anticipation, and how the heartless flirt had tantalized him with false hopes and at last had cruelly deserted him—flung him off: and then he told us that his eye-teeth were now cut. Ah! brethren, brethren, how many of us have had our poor pates lured into this same false Delila's lap, and have been deceitfully shorn of our precious locks, and awoke only to find our former strength departed; and, alas! how many of our noblest sires, too, like yours, Mr. President, frosted with the experience of many crops, have been captivated by her smiles and made her willing dupes.

'Tis said that "there is a tide in the affairs of men, which, when taken at the flood, leads on to fortune." That fortune, my brethren, has not ebbed in on the flood-tide of cotton which is surely drifting us into poverty.

But it is human to err. Let us give over the wild chase—cease to follow the will-o'-the-wisp; let us go back to good old Uncle Corn once more, and to our more reliable country cousins, the Misses Oats, Peas and Pumpkins. We know their friendship; 'tis tried and true.

Look around us and behold the common wreck. Debt and bankruptcy are sinking the hearts of men into the dark and turbid waters of despondency. Where are the honest, jovial faces we were wont to see in days of yore? Gone glimmering among the things that were, and in their stead we see, at every turn, the longated visage, the downcast eye, and the pendant under-jaw. Ask for the trouble and they will tell you the old and too familiar story—had an attack of cotton on the brain. The awful epidemic had seized them, like some thousand-legged nightmare, stagnated their blood and, like grim death, pinned them down, and the future offered no hope. But occasionally you meet a contented face. Ask how so—how have you escaped the general ruins—and he will answer: "Well, sir, I raise my home supplies; I never go in debt; every year I make a little above my own needs, and to these fellows who raise all cotton, why I sell them something to eat, sir." And hereby hangs a tale.

Let the farmer give his first attention to home supplies, fill his home with comforts and contentment, then let the chords that support the Wall street rigging snap asunder. Let the main masts and money kings topple and tumble; let financial panics and crises come. Amidst the crash, the self-sustaining farmer will float serene; with barn and store-house well filled, he can snap his fingers and whistle Dixie.

There is a terrible hydra-headed monster on the rampage throughout our land. A merciless dragon of consumption, his trail is marked with wan despair, and like a besom of destruction, he sweeps the country. His name is debt. The people know him, fear and tremble in his presence, yet madly rush into his very track. Loans and liens are his daily diet. The ever insatiate beast, with hungry jaws crammed with cotton bags, still cries for more and more; and his infatuated victims hurl the overburdened commodity into his throat—and are frequently swallowed up themselves. Is there no deliverance? Yes, thank God, a few wise men have seen a star. A saviour has been found; an angelic song has been heard, proclaiming peace and good will to the tiller of the soil. His name is Cash, and common sense catches up the strain, and chimes along the farms, pay up, pay up as you go.

Mr. President, I believe it is the farmer's true policy, if he can't run ten plows on the cash salvation plan, to come down to five, or two, or even one; and if he can't make the riffle with one, then to quit the business, or hire out to some man or woman who does business on that plan. If he can't work fifty acres well, then ten; if he can't pay cash for his fertilizers, then save what he can from his barn-yard, plow deeper and cultivate better. If the Dixons and Wothens can make from three to five bales per acre, why should we put up with one bale for from three to five acres? Brother Mulligan said muscle and brain were needed, and he is right about it. The fault is with us. If our patches are not just what we want them to be, we must make them so. Our Creator has done his part, and left it to man to develop the hidden resources stored away in nature's labyrinthean recesses; earth, air and water, all are teeming with material to supply the wants of man.

Necessity has been called the mother of invention, and the direst necessity often produces the most beneficial results; and who knows, brethren, but that the very difficulties which now encompass us may be fraught with some great blessing to the tillers of the soil. We have gotten into a fog; we must arouse to a sense of our danger, and with strong hands steer clear of the disastrous rocks of debt, too much cotton, and poor culture.

Mr. President, I believe more profit can be realized from ten acres well cultivated than from fifty in the ordinary way, thereby both lessening the cost of production and increasing the profits of the farm, besides the improvement of the property; and if my proposition is correct, then cash and high culture are the true finger-boards to successful farming, as all will agree. Why not adopt the plan at once? There's the rub. How to get at it is the thing. Some are so deeply in debt that they think they can not adopt the cash plan; and so many a poor sinner wants to believe in the Saviour, but hesitates to lay hold on the salvation plan, still delays and tries to work himself into a more acceptable state with his God, but only succeeds in heaping sin upon sin on his poor soul. And the planter, in trying to get out of debt by going in debt, is getting in deeper all the while. The present southern farmer has to be regenerated—to be born again—to go to his creditor, like the sinner does to his Saviour, give up all he has, if necessary, and start a new and better life.

I believe, sir, farming can be made to pay; I think we have cause for encouragement if we can profit by past experience, and appreciate the lights before us; and what avocation is there in life more desirable than farming, what occupation can afford more attractions, what more free and independent, and where on earth ought woman, the true wife and mother, to find more real happiness, where more contentment than as mistress of some good farmers household?

The farmers make a great mistake when they select their dunces for the plow handles; they should pick their brightest boys for the farm, and put the fools somewhere else. They may fill some other place, but the farm never. It requires as much brain to conduct the farm successfully as it does to legislate in the halls of congress. A farmer ought to understand all the requirements and deficiences of his soil—to be familiar with the agricultural experience and improvements of the world. He ought to be an expert even on the rostrum, for I believe the time is not far distant when he will control this great government. He holds the balance of power in his ballot, has the biggest share of brains, and only needs the culture. Cultivate him and he will take his true position in the world, and then he will frame laws to protect himself, and advance the cause of agriculture, and wrest from the cormonants of the country his rights, which so long have been trampled upon, drive the money grabbers from their high places, and save the people from the avaracious craws of the few, and then the laboring masses will get their dues.

Mr. President, I feel like I believe it will not be long till the daylight will begin to dawn on the tiller of the soil. I believe the farmers of this country will rise in their might, and claim their own. I think I can discern the first rays of the morning light, the herald of the coming sunshine, and if we live a few years longer, Mr. President, we may see it rise in its radiant glory, and our chidren may see it ascend higher in the horizon of intelligence to its noon-tide splendor, till its fructifying influence shall make the world better and happier. But a short time since we first heard of the grange, and even now the name is scarcely familiar to our ears. Their power unknown to themselves—a power, though in its infantile experience and ignorance, that is yet shaking with the sound of its voice the very heart of this corrupt government. May God give wisdom to the laborers and grant that their combination and honest efforts may prove an ocean of blessings to this country. I know, sir, we must expect our share of the ills of life. Darkness has hovered over our recent pathway, but I believe if we will have it so, there is a better day coming, bless the Lord. Let us only be true to ourselves, and we shall bring the world to our feet. We have the elements of power; let us cultivate the brain.

Now a word in conclusion. I maintain that a farmer, a granger, cannot fulfill his true position in life if his aspirations are no higher than to grasp the perishable of this world. Man at best is but a pilgrim upon earth, and but for a season; is on a wearisome and hazardous journey, and if he will but cast his eyes beyond "this vale of tears," he will find it is "not all of life to live or all of death to die." Solomon tried it all, and concluded that there was very little here besides vanity, and about the best thing a man could do was to eat and drink what God allowed him,

Now let the farmer heed the commands of a merciful God, and strike only for his just rights, which he has not by many jugs full at present, abandon all inordinate desire for greed and gain, and his home may be made, indeed, a place of contentment, where he may sing, "home, sweet home," and where it may be felt that there is "no place like home," and when done with the earth, our voices may be attuned to a higher sphere, where we may join the heavenly choristers in the everlasting home of homes. Mr. President and brethren of the plow, I offer these thoughts, and home supplies in abundance recommend the cash plan, high culture both of land and brains stand shoulder to shoulder, live in the fear and admonition of the Lord, and I guarantee success in this world, and a far better country in the dim mists of the eternal future.

PROFESSOR N. F. COOLEDGE,

A distinguished educator; born in Vermont; came to Georgia a young man; first taught school at Perry, Ga.; became famous as a teacher at Cotton Hill, Ga.; afterwards taught at Dalton, Canton and Norcross, Ga.; still resides at the latter place, retired from business; father of the Cooledge Bro's., of Atlanta, and an earnest working Baptist; has been one of Georgia's best teachers.

THE UNEXPECTED PREACH.

At the time of this story, our home was in Norcross, Ga., on the R & D. R. R. One of the most venerable and useful citizens of this town was Professor N. F. Cooledge, a distinguished educator, and an earnest Baptist; a fine, portly looking old gentleman, and one whose appearance would attract attention anywhere.

The professor and I made a trip to Cumming, Ga., up near the mountains; spent the night there, and hearing of a Baptist camp-meeting across the Swanee mountain, concluded we would attend on the morrow, which was the Sabbath. So in the morning we ordered out our conveyance and drove over. Arriving at the enterance to the camp-grounds, we were met by several clever looking countrymen, who had our horse cared for and bestowed on us, as we thought, extraordinary hospitilaties. We were invited down to the stand, as it was about time for the morning services. Instead of entering the aisle at the front, we were conducted round to the rear, and before we were aware of the situation, were being ushered up into the pulpit. We remonstrated, but they persisted, and introduced us to the preacher, who had just risen to start the opening hymn. We were seated, one on the right and the other on the left on the preachers' bench, and left, to our own reflections. As Brother Pirkle proceeded to line out his hymn a sudden idea struck me that Professor Cooledge had been taken for a preacher, and would be called on to follow Brother Pirkle. It was too good, and tickled me all over. I knew the professor would be astonished, taken completely by surprise, and in my imagination, I could see his eyes about the size of saucers—had to pinch my thighs severely to keep from laughing outright.

Brother Pirkle finished, and turning to me, said; "Brother Sloan, you will follow me." This shock came as a thunderbolt from a clear sky. Such a turn of affairs had never entered my calculations. I half arose, completely befuddled, saying, "No, sir; I—I—I—," but the preacher paid no attention to my remark, and said, "let us all pray," I knelt at my seat, but didn't take in much of that prayer. My great desire was to escape. I twisted round to get a view of the long steps we had just come up. Every plank was packed with people, and found that I would have to make a leap of at least fifteen feet to clear their heads, to get to the woods. I raised up high enough to peep over the pulpit to the rear, but there were rows of heads. At last Brother Pirkle said amen, and arose to read and announce his text. I tried to attract his attention, but in vain; my tongue seemed to be paralyzed, and I felt as if what little sense I had ever claimed, had departed. I did not look towards Professor Cooledge, partly from a guilty conscience, and did not care to catch the expression of pity and anxiety I knew to be upon his face, in my behalf. I was in a predicament, and how to get out of it I could not see. In my checkered life, I had faced many dilemmas and dangers. I had been among wild Indians, chased on the plains by wolves, on a burning ship in the midst of the rolling billows, and had passed through the terrible carnage of war. I remembered in my school-boy days the relentless rod of the pedagogue, the agony of early manhood, my disappointment in love; had passed through an unusual share of the ups and downs of life, made bad speculations, been dead broke, but in the whole category of my tribulations, I could think of nothing so embarrassing as this—for me to preach to four thousand people, and me not a preacher. There seemed no way out of it but confusion and absolute disgrace. To preach to a camp-meeting, and already scared most to death! My heart thumped against my side, my legs were all in a tremble, and I felt a great weak strain down my back bone. Time was passing, and the crisis approaching. An old saying flashed across my mind, that a cornered rat will fight a cat. Then I thought of old Preacher Dannelly of South Carolina, the most self-possessed and confident looking man I ever saw, except Sam Jones. I perused the pluck of both of these men, and it helped me. I knew something " had to be did," and concluded that the best way out of the ordeal was to wade right through the fire. I resolved to try it; reached up got a hymn book, selected a number and turned down the leaf, listened attentively to the sermon, and marked in my memory some of the principal points. After a time the preacher finished, and turning to me, waved his hand to the front. I arose as deliberately as my shaky legs and yielding back would allow, and leaning with my right arm on the book-board, the book in my left hand, and for a moment surveyed the sea of heads around me, then proceeded to line out the song. When finished, all excitement had vanished, and I entered the skirmish line without a single feeling of my former terror. I complimented the able sermon of the preacher, commented upon the unanswerable points of his argument, extenuated upon the great truths advanced by him to the dying sinner, and closed my ten minutes talk, with an army anecdote, applicable to salvation, then turning to my left, in a most solemn tone, called on Brother Cooledge, to lead in prayer, to which he responded in the most efficient manner. The preacher closed the service, and we were invited to a tent to dinner. While sitting at the table, a couple of committeemen came in and announced that Brother Sloan had been appointed to preach the evening sermon. Then I squealed, and let the cat out of the wallet; told how I had been taken in, and that I was no preacher at all. Of course, I was excused.

The professor and I started home after dinner. We had ridden along some distance in silence when I remarked, "Professor, I believe you are about as deep in the mire as I am in the mud; suppose we don't say anything about this scrape when we get home." He said he'd never breathe it.

About a year afterward, two young lawyers came down from Cumming and stopped at the Norcross Hotel. I was proprietor, and was carving at the dinner table. At the table also, were quite a number of Atlanta guests. The two young lawyers seemed to be having a side-show of fun to themselves. When I asked them to divide, give the public the benefit of their mirth; they asked if I really wanted to hear the joke. I told them by all means let us have it, when George Bell waved his hand and said, "Brother Sloan, you will follow me," and then blurted out the whole story; and to make matters worse, my wife remarked that they must have kept the matter very quiet, as she had never heard of it before. I had to grin and bear it all, and became more than ever impressed with the old adage "that murder will out."

A HISTORIC HORN.

From the Constitution: "Mr. D. U. Sloan, of the National Hotel, has a historic horn, and on being asked the story connected with it, furnished the following sketch:

"This horn has been in my possession for one-third of a century. Notice the perforations through its rim; see how the worms have eaten it. It was presented to me by a man I never saw, nor heard of in my life until after his death, and who never saw or heard of me. His name was Kirkpatrick, and it came about in this way: Kirkpatrick was on his death-bed, and said to his friend Strohecker, of Charleston, who was sitting by his side: 'Strohecker, there hangs a horn. I have prized it much, on account of its superior tone. The delights of the chase are all over with me. I shall never be able to sound it again. Take it, and give it to some good hunter, for me, and tell him I bequeathed it to him as a dying gift.' Strohecker promised, and I became the favored one; and if departed spirits have cognizance of what happens here below, I trust the old hunter may be satisfied with his legatee.

"I have winded this old horn in many a hunt on the Blue Ridge mountains, with the Hamptons, Calhouns, Haskells, Taylors, and many others of South Carolina's noblest sons. I made old Charleston's walls ring with its shrillest notes, on that memorable evening of secession. I sounded it again, on Atlanta's hills, for Cleveland and democratic victory, and made it to resound with lusty blasts on the triumphal entry of Jefferson Davis into Atlanta. I was a secessionist, and fought for what I believed to be the rights of my country; and though a reconstructed rebel, I do not feel that I committed treason against the federal government. If so, our fathers of the revolution did the same thing. The same causes existed, but God gave success to the one and defeat to the other. His ways are inscrutabe, and we know 'he doeth all things well.'

"The lost cause is dead and buried. I revere its ashes, and love and honor the grand old chieftain, who must soon go, too. I honor the old hero, because he never faltered, nor shrunk from what he believed to be his duty.

"But about this dear old horn. I shall hope to sound it again in 1888, for Grover Cleveland, or some other democratic president; and if defeat should be our fate, will hang it among the willows for another and more propitious day. Once before then, however, I will take it down and give three blasts for our next governor, John B. Gordon—a name irresistible to every son of Georgia, and to every boy who wore the grey. Respectfully,

"D. U. SLOAN."

"Atlanta, Ga., May 30th, 1886."

From the Atlanta Capitol: "This morning a Capitol reporter stumbled upon an item that will be read with interest, and will also be amusing. It will be remembered that Mr. D. U. Stone, of the National Hotel, has a historic horn, which he sounds out on patriotic occasions. A communication appeared in the Constitution, about the first of the gubernatorial campaign, in which Mr. Sloan wound up by saying he would sound three blasts from his horn for Governor John B. Gordon. He has received the following postal from a Bacon man:

"'ATLANTA, GA., June 6th, 1886.
' Mr. D. U. SLOAN:

'Dear Sir—I have read with much interest the account of your historic horn, but would suggest that you practice on it from the reverse end, as you will have to blow it out of the little end, for Gordon, when the convention meets. As I am a private citizen, and have no axe to grind, have clipped my name from this card.

'Yours truly, ——————,'

'Sloan's reply, through the Constitution:

'MY DEAR UNKNOWN FRIEND—Your card with name clipped off is received. I read and considered its contents, and thought, 'Is it possible that I am mistaken; shall I, indeed, ever blow this good old horn out of the wrong end for John B. Gordon, the soldier, the statesman, the people's man?' While thus sadly ruminating, I seemed to hear a voice—a whispered voice. I turned and listened. The old horn was trying to talk, as it hung above my head. With bated breath I listened, and these are the words I caught: 'B-y-e-g-o-n-e, b-e-g-o-n-e, B-a-k-e-o-n.' I arose and reversed the ends—turned the right end, the mouth-piece, to the breeze that played through my open window—and the words changed and these are the sounds I heard: 'G-o-o-d-o-n-e, g-o-o n, G-i-d-e-o-n;' and as a stiffer breeze struck the right end, it spake out distinctly, 'G-O R-D-O-N, G-O-R-D-O-N, G-O-R-D-O-N.' So, my dear unknown friend, do not allow yourself to be deceived. This is not only an historic but a prophetic horn; for even as your name was clipped from your erring card, so shall the wings of your aspirant be clipped of his expectant glory, when the convention meets, for, most certainly, I shall sound the three prophetic blasts for Gov. Gordon.

'Respectfully, D. U. SLOAN.'"

From the Atlanta Journal:

"During the immense cheering, and great excitement, in the gubernatorial convention, attendant on the nomination of General Gordon, there rose high above all the noise and din, three sharp clarion notes from Sloan's historic horn. In a moment a dead silence reigned for a brief period, and was broken by a voice, shouting, 'That's Sloan's horn; toot her again!' then the cheering was resumed with a will "

DRIED APPLE CIDER.

IN a previous chapter, I stated that I had long been impressed with the idea that I was a born speculator, and although my experience in life had been sufficiently disastrous to entirely explode this pet theory to any ordinary practical person: yet, I still condoned my constant reverses with the excuse that I had not struck it right—had not struck the ebb at the flood-tide that led on to fortune, and with unbroken spirit still looked hopefully and fondly to the future, when things would turn up more favorably, and even now, seemed the auspicious time, and, indeed, in this dried-apple business, things did turn up mightily, but not in accordance with my pleasureable anticipations, and turned up with such dynamic force, as to greatly shake my life-time faith as to my birth-right as a speculator.

When the great prohibition movement resulted in success, I was proprietor of the National Hotel, and one of my frequent guests and warm friends was a Mr Obediah, who owned a fine river farm near Gainsville, Ga. There he cultivated big apple orchards and vineyards, and manufactured oceans of vinegar, and sold profitably to the various markets·

One day Mr. Obe registered at the desk, and I noticed a peculiar cunning twinkle about his eye, and soon he had me off to one side and was divulging a great scheme— the result of much figuring and meditation—an enterprise, the manufacture and sale of cider. Prohibition had now become a sealed fact; now was the opportune time; the people couldn't get whiskey nor beer to drink, and consequently would take powerfully to cider. I asked where the apples were to come from, at this season, to make the cider. He gave me a knowing wink, and answered, "Dried apples; the best cider in the world; equal to champagne." He had recently bought a recipe at an extravagant price, which would keep the cider sweet indefinitely. Said it would be the biggest business out; showed the immense profit to be made, and said he had selected me, as the man he could trust, for his Atlanta partner. As he unfolded his well matured plans, I saw every thing plainly, and even more, too, than he had yet conceived. The firm was organized, and the duties of each fully agreed and understood. Mr. Obe. would furnish the barrels and kegs, and manufacture and ship me the cider; we would quietly buy up all the dried apples on the markets, and empty bottles; I to provide delivery wagons, and the necessary help for the sale of the cider. Our plans all arranged, Mr. Obe. returned home to manufacture, and I to prepare for the sale and delivery. The first thing, I found a large quantity of dried apples at Mr. Shomo's, bought and shipped them to the factory; then cleaned the city out of empty bottles, both pints and quarts, but met with a loss on the pints, as the law would only allow us to use the quart bottles; rented the back end of Cohens store, on Alabama street, and the privilege of an ice-house, for storing; got up a delivery wagon, and made engagements for sales. Everything worked nicely, and I had confidently considered the question of many investments in Atlanta dirt. I sent Mr. Obe. word to turn on a sluice of dried apple cider—that all was ready—and promptly received a cargo of barreled cider, and stowed it away in the ice-house; hired help and bottled up a couple of thousand. Mr. Obe. came down to see the business well started, and we loaded up the wagon with the bottles in boxes prepared for the purpose, and a keg which had been engaged, and then mounted the spring-seat, and moved off. The business was in operation; we delivered a dozen bottles here and two dozen there, and the keg, according to engagement; and as we traveled round delivering, were in charming good humor, and very much in love with each other, and all the rest of mankind.

We were moving far up Decatur street—the day well advanced and the sun growing intensely hot—when we heard a shot in the rear. We turned to see where the shot came from, when "Bang!" went another, and a cork flew over our heads, with a shower of cider. This exhibition had not been put down in our original programme. We considered it accidental, knowing that accidents sometimes happen in the best regulated families. Stopping in front of a grocer's store, Mr. Obe. stayed with the team while I gathered an armfull of bottles and went in. I found the proprietor and family in the back room at dinner. I made them a little speech on the merits of our champagne cider, and remarked that it was a nice opportunity to give them a taste of our delicious beverage. I cut a wire and, before I expected, the stopper and the foaming liquid burst out and struck the old lady full in the face. I turned the muzzle as quickly as possible, and it bespattered the bosom of the daughter; whirled the gun from her, and the old man, in trying to dodge, turned his chair over and fell sprawling on the floor. The ladies fled, screaming—and the old man cursing. I was left alone in an empty room, with an empty bottle. I tried to follow, to apologize and explain, but they shouted at me, "Get out, get out; take the derned stuff out!" When I got to the front door, I heard several more bottles firing off, and Mr. Obe. was swinging to the lines to keep the horse from running away. I climbed in behind, and we started for home; and as we pranced down Decatur street the fusilade opened out in dead earnest, and it took both of us to keep the team in the street. And the people in the streets, doors and windows, gazed in wonder on the passing scene. We got safely back to the store, and found all in confusion and consternation there. The bottles were firing off in platoons in the rear end, the corks striking the ceiling and flying all over the room, and the inmates huddled about the front door. We stood in speechless horror at the scene. Just then, the man we had delivered the keg of dried apple cider to, came rushing up and reported that the keg had blown up and torn the whole side out of his house. Cohen was ranting, and wanted the dynamite removed from his house immediately; but the demand was unreasonable, and we paid no attention to it. No man could be had to face that terrible battery. Somebody suggested Cap Joyner and the fire department, but Cap could do nothing there. Some wanted Connolly and the police, but several policemen peeped in the door and then shied off.

After awhile, the fracas gradually exhausted itself and then died down, and was succeeded by the usual calm that follows the storm.

When some new customers came in (who had not heard of the trouble), inquiring for the champage cider, we took them down to the ice house and tapped a barrel with a mallet, when the bung flew out like a cannon ball and sent a fountain of cider drenching the party, and everybody fled from the scene. Other explosions followed till everything was empty.

Mr. Obe. and I dissolved the firm by mutual, silent consent. He resumed the manufacture of vinegar, and I confined my efforts strictly to the affairs of the National Hotel.

But I have since thought we broke ranks prematurely, and lost a great opportunity, one that might have proved a fortune to us, as the power from that dried apple cider might have been most profitably utilized (instead of the engine) under the artesian well. Why, there was force enough in one of those kegs of dried apple cider to have thrown the water clear over the Kimball House, and rushed it through the piping to every part of the city.

AN OLDEN TIME FOX CHASE.

The people of to day have a greater variety of amusements, than in the olden times, and I suppose their amusements must be attractive to them; but I wouldn't give one good old time fox hunt for all of theirs bunched up into one big show. As to their germans, I can't form an opinion, for I never saw one. The base ball I dont understand; think the old town ball is good enough. As to their clubs and secret societies, I care nothing about them; I dont like the secret business. When I get hold of anything good, I want everybody to know all about it. The modern circus has got so many rings running at the same time, I can't see what is going on in one, for being bothered with the others; and even music is now so adulterated and diluted with cranky preludes, and foreign variations, innovations, combinations and complications that it is hard to detect a bit of the old simon-pure in it. And now they have got to having canine exhibitions on the stage. (The theater has gone to the dogs sooner than I expected.) Recently a dog-gone professor introduced a parcel of imported whelps on the stage in Atlanta. He had along with him a vagabond Irish dog he calls Barney, that stood on his head, and the people thought it wonderful. I would like to know what use, or common sense, or skill there is in a dog standing on his head. A good sensible dog, in our day, would have refused to have made such a fool of himself. Then this professor of dogs had these dude poodles dressed up in silks and streaming ribbons, parading the streets, drawn by splendid spans of horses, in magnificent carriages. Think of it, American people! Dogs in silks, dogs in carriages, and dogs on the stage.

I thought it bad enough to try to " histe " the nigger over the heads of the white folks, but now it comes to " histing " the dogs over both—what next? I like the dog and I like the nigger, but I like them in their places. I like rich folks and poor folks, but there is a proper place for all. But it does look like, in these modern days, things are getting "sorter mixed;" but for real useful knowledge and intelligence, we had dogs just as far ahead of these gentry imported pups as Thomas Jefferson was, in his day, in true, broad statesmanship, ahead of little Benny Harrison. This seems to me to be a day of queer capers, anyhow. Just think of Clay, Calhoun, or Webster cutting up the capers of Tom Reed in the last congress. Why, it is too ridiculous to think about, and these gentry dogs seem to be running in the same line.

My memory goes back to the days of old Troup, Hector, Baily, Rattler, Jeff, Lady, and Haidee, and other good dogs of their time and kind. Dogs, in their day, noted for their dignity of character, their unquestioned veracity, their almost unerring wisdom in the science of trackography, their vast attainments in deer and foxology, dogs of sterling integrity, who deserved to be, and were, examples, and were imitated by every respectable young dog and puppy that came within their purview. I have often watched these old dogs, as they lay down, or squatted in the summer's shade, meditating upon the mistakes of the last season's hunt, and planning to avoid all such in the next. I have seen them so absorbed in such reflections, that they would forget to snap at the flies swarming round their heads. I have seen old Troupe go off into a snooze and get to dreaming. Sometimes he would dream he'd struck a fresh, hot trail, when he'd spring to his feet and shout out, "h-e-r-e h-e--w-e-n-t," wake up all the balance of the pack, and have them charging around looking for the game; then he would look ashamed of himself, walk off sorter grinning like, hunt another place, and lie down again.

If Troup thought he knew where he could start a buck, or wanted to go a hunting, he would come to me and whine and frisk, and wag his tail, and look off toward the mountains, in the direction he thought the deer was; and if I couldn't go, I'd just tell him so, and then he would look disappointed, and if he felt he couldn't stand it, he'd go and wake up the pack; and if they were too lazy, why he'd just go by himself, pick out a good sized buck and run him clean to water.

Troup was a philosopher and an economist. If he thought he was going to have a long run, he would economise his wind —he'd only open about every quarter of a mile, just enough to let it be known he was coming down to his business. This old dog caught in the Blue Ridge mountains sixteen deer, that never had a shot-hole in them; and old Jeff broke one of his fore legs running a deer. I splinted it up, and he went out again and broke the other leg, and walked home three miles on his hind legs, and for months he walked about the yard on those two legs. It's a fact. My wife says she has seen him do it a many a time, and she will tell anybody so.

But we started to tell about an olden time fox chase. We have been in so many—hardly know which one to tell about: Our old time dogs couldn't speak English, but they could listen, and heard every word we said, and knew just what we said, and what we wanted.

I'll give this one. We were to meet at Warr's old field, which lay between the present town of Seneca, S. C., and the river, at half past 10, o'clock p. m. I select this one because it was short and business like.

As the hour approached, I mounted my Bucephalus, and blew up the dogs, they were all keen, in for the hunt, and in finest trim. As I wended my way along the country roads, I could hear the winding of the horns of my comrades as the cheery sounds were wafted over the hills from their different routes. We were all promptly on the ground. There were Tom Lewis, Dave, Mack, Sloan, Joel Patterson, and myself—all had our favorite dogs. We were sitting on our horses discussing the route to be taken; the dogs were flying round in wide circles over the crisp grass and frosted leaves, in the bright moon-light. Joel Patterson had a dew drop in his pocket, with which we all moistened our lips. Joel had hardly returned the jewel to his pocket, when Tom Lewis' little bitch, Lady, struck a trail, down near the old school house. Tom yelled out, "I heard you my little Lady," and we all moved off in that direction. We passed old Jeff in the field, (close on our right, diligently snuffing the ground,) and saw him suddenly raise his head heavenward, and quivering from head to foot, cried out, in doggerel :

"He's been right here, for a fact; he's been right here, I smell his track."

Now Rattler opens out across the branch, and the cry becomes general, as each dog gets a sniff of where reynard has been loitering. Now Hector and Bailey turn loose together, down the ridge; they have got the running trail; they make towards the fish trap, and the whole pack have fallen into line, and the music has commenced. Talk about Gilmore's, or Barrack's bands, I love to hear them; but if both of them were going on at Grant's park, and this old time dog music heard about the Soldiers' Home, I'd make for the home. Now every thing is in, and its " Yah-rah-ya-rah-ugh-

ugh-oph-oph-ya-rah-oph-oph-ugh-ugh," and notes impossible to spell in the English Language, or for the science of music to confine to it its limited staff and bars. Now the chorus swells out o'er hill and dale, with its prolonged and softened echoes, a music wild, wierd and heavenly. I imagine it to be a foretaste, a sort of type of when, in the millenium, the angels shall come down to chase the devils out of a sin freed world. Oh, the extacies of an old-time fox hunt.

The dogs are off, and the hunters close behind, yelling encouragement to the eager pack, leaping logs, gullies and fences. Now they turn up the banks of the Keowee, and now their quickened cry tells us plainly they have got him on the run. Now they turn again at the fork, and up Little River, and now they leave the stream and make for the Dry Pond, and on toward the Ramsey place; there they turn up Seneca creek, cross over and down the other side. We cross the creek and wait; here comes the fox in a few feet of us, he bounces like an india rubber ball, seemingly confident, and at each jump his tail flies from the ground high over his back; the dogs pass us well bunched, and wild with the excitement, as they see us watching them, and five lusty throats gave the genuine foxwhoop with a will; now they head round Sloan's mill-pond, and on down to the Earle place; here reynard tries several dodges, but the dogs push him too close, and he resorts to the cow-pen trick—in amongst the cattle. As the dogs come up the cows bellow and show fight. This brings confusion for a little while, but the dogs circle wide to avoid the cattle, and soon Heck and Bailey strike the trail again, and all the balance of the pack fall quickly into line. Now back to the mill pond, where he tries to loose them in the hurricane thicket, but they soon scare him out, and back to the Ramsey place, where he tries the fence-dodge; but no go, for Haidee finds where he struck ground, and off the pack go, now back to Warr's old field where the start was made. Here he makes several circuits. Now we see him again, badly worn and his tail hangs low; he is running for his life; the dogs pass us tired but confident; now he makes for the fish-trap once more, but turns as we meet him again. We see his time is almost up; his efforts are labored, his tongue is lolling from his mouth, and the fog is rising from his thick fur; the dogs are fast closing the distance between them; we follow close behind, and now old Bailey's nose is almost on his tail; he turns to avoid the dog, and falls into Rattler's jaws, then a bunching of dogs, a scramble, a death squall, and all is over. Bale Maxwell gets the tail, the fun is over, the hunters are happy, the dogs are happy, and poor reynard, if not happy, is at rest. The dogs rest, loll their tongues and pant, we blow our horses and talk over the chase, and the splendid performance of each dog is commented upon. Now we wind our horns again, hunters and dogs separate and seek our several homes to sleep, and dream happy dreams, and as we write up this old time fox chase, after the changes and tribulations of time, the whole of life seems to us but a fitful dream.

We are aware that some men are soulless as to the music of a fox chase, for we remember Bob Jarrett, of Tugaloo, once called out an infidel to listen to the glorious melodies of a fox chase. The dogs were in full cry up the river bank. Bob asked him again, "Dont you hear the glorious music?" when he, the idiot replied, he could not hear a thing for the barking of those confounded dogs down about the river. Bob left the man in unutterable disgust, and sprang out after the dogs.

HON. JONATHAN NORCROSS,

Born in Charleston, Maine, his first adventure was in Cuba, where he engaged in the machine business; next in North Carolina, as a school teacher; from thence to Augusta, Ga., in the same avocation; thence to Marthasville, with a horse saw mill; next, established at the now famous old Norcross Corner as a merchant (where I well remember his shingle advertisement hung out, for exchange in country produce).

Elected Mayor of Atlanta in 1850, Mr. Jonathan Norcross may be considered above all others, as a father to Atlanta, as a father to the great Air Line Railroad, and the father of the State Railroad Commission. The Pioneer of Atlanta, and as the living link between its present and the past; has lived in Atlanta from its beginning, and its friend in every great enterprise. An honest republican, loved and honored even by his political foes. I have known him for forty years, and remember him in those olden days as one of the best patrons of the telegraph business.

THE CRACKER GIRL.

On Carolina's hills, my father's flocks
    Were fed, and I a mountain sprout,
When wandring fancies filled my head,
    And I longed to look, look about.

From South Carolina, a frugal swain,
    Like Norval from the Grampian hills;
Though we left home in times of peace,
    Our home where sang the whippoorwills.

Had heard tell of Iron rail-roads,
    Big towns, of steamers on the sea;
Of a great world filled with wonders,
    So a rov-yer were bound to be.

Had read in books about Columbus,
    Of Alladdin, in the Arabian Knights,
Of clever old Robinson Crusoe,
    And of strange and marvelous sights.

It was talked, away over in Georgia,
    A mighty town had been designed,
'Twas bound to be a railroad center,
    Oodles of trade would be consigned.

Marthasville the little burg was called,
    Now as Atlanta was to be known,
Taken on a high-fa-luten name,
    Had let out a tuck in her gown.

Now Martha was but a country lass,
    And so she'd stepped upon the rink,
Although her dress was sort of shabby,
    'Twas said she had a business wink.

And so I left my parental home
    And steered my charger west-ward ho?
My fortune in my britches pocket,
    Lit out, struck the grit, off did go

For Atlanta, in a bee-line course.
    Here, forty years ago, I landed.
'Twas not long till my funds were gone,
    Till my finances all were stranded.

When I found I'd have to hump it;
    Fortune favored in my behalf,
A new thing had just then started,
    And I struck for the telegraph.

To President Foote I boldly went,
    Applied for the Atlanta position.
He asked if I was an expert,
    I said, to be, was my intention.

Told him I'd never seen the thing,
    But reckoned a fellow might learn.
Astonishment seemed on his face,
    Tho't he said, well, you-be dern.

I told him I wan't afraid of work,
    Then looked him right straight in the face.
Said he, young man, may be, you can;
    Go try, if so, can have the place.

I'll ne'er forget his kindly face,
    When as learner I was installed;
I made the riffle, caught on the lick,
    By no obstacles was appalled.

First to sling Atlanta's lightning,
    'Twas right here I made the start;
Then but a little shabby hamlet,
    And now this great business mart.

The town then seemed a small potatoe,
    A sort of grass colt over done,
And so slow to grow any bigger,
    As to outcome there seemed none.

And ever since that's been the talk,
    Folks said she'd never fil. her gown,
But she's kept a growing all the same,
    Till the old dudds can't hold the town.

She's bound to be a corset burster—
    Can't tell what she's going to be;
Of all towns in the Sunny South,
    She's bound to be the grand Cit-tee.

But I got tired of the cracker girl,
    And longed to see the far off west;
Now for California gold mines,
    There next determined to invest.

So with my friend, J. W. Rucker,
    We bade the little town adieu;
Agreed to try the world together,
    Those auriferous fields we'd view.

We passed within the golden gate,
    But it cost us many a quarter;
Found that gold did not grow on trees,
    Had to be dug in mud and water.

Sure we took in those golden fields,
    Didn't pan out as we expected;

J. W. RUCKER, Esq.,

Of the present firm of Maddox & Rucker, Atlanta, Ga., is my old California chum. Forty years ago he was a poor young man, a clerk for U. L. Wright, on Whitehall street. We left Atlanta together, went by my father's home at old Pendleton, S. C., from thence via Charleston, S. C., New York, Havana, Panama, Acapulco, San Francisco, Sacramento, and into the gold mining region, where we soon got strapped and had to separate and hire out to work for a living, and met again in Atlanta, Ga., after many years.

Rucker, by great diligence and unswerving rectitude, has accumulated a handsome fortune, is well known in Georgia, and liked by everybody.

We had no lack of hardy toil,
    But as to wealth, wan't elected.

'Twas then we thought of home, sweet home,
    Of friends and comforts left behind;
Of our dear old South and betterments,
    And of many other things in kind.

I sighed for the songs of whippoorwills,
    And longed to herd my father's flocks;
To see the dear old hills of Caroline,
    Divel take the pesky gold and rocks.

Rucker yearned for the cracker girl,
    Sweet to him ee'n in her shabby frock,
Swore if he got back to her again,
    There he'd forever plant his stock.

True to his word here he has staid,
    Has proved most faithful to that vow;
Plodded on through the weary years,
    Till his form has begun to bow.

I went back to old Caroline,
    After years wandered here again;
Could not forget the cracker girl,
    Who fixed her image on my brain.

Found Rucker on an upper limb,
    Though I am roosting on the fence;
I am glad to hear Rucker crow,
    He's got dollars, where I've got cents.

Rucker was sort of slow and sure,
    I, perhaps, little sorter fickle;
But he's raked in the spondulicks,
    And I've just about lost the sickle.

Right here will tell of another friend,
    A friend of all, would do to trust:
L. E. Bleckley, who'd settled here;
    Anchored here, for better or for worst.

Friend Bleckley bought a box of books
    At auction, said he'd no time for fun;
To read and learn, he'd started out
    For to climb the ladder, he'd begun.

Then, too, we had a debating club,
    And he was the longest winded,
On questions for to argufy,
    With fellers that was so minded.

Bleckley kept a mighty digging,
    Digging for lore, law and fame;
I didn't take to that kind of digging,
    But he dug, dug himself a name.

Now I'll speak about a little boy,
    Little black-eyed kid, name of Evan--
Evan Howell, my dispatch boy,
    And this kid is still a liven.

He's a big horse now, a rouser,
    Is printen of a great big paper;
He don't seem like that small kid now,
    Case he's a real gol-golly whopper.

Recall to mind many others
    We left here forty years ago;
Most of them have gathered moss,
    As their Atlanta dirt will show.

Had we stuck to the little city,
    And, like them, had saved our pewter,
Might have become a plowshare too,
    Instead of a little scooter.

> This page was intended to present a likeness of my friend, Judge Bleckley, but at his request it is left blank, so that each reader may supply the picture from his own imagination.

The lines above were written and handed to me by my friend, Judge Bleckley, upon my application for his picture, and it is with regret that I have to inform my readers, that on account of my regard for the request of the Judge, I have restrained my desire to present his likeness on "this page," as intended.

However, as the Judge's injunction only applies to "this page," as intended, it is with pleasure that I announce to them that I have secured a most excellent picture of the Judge, and present it on another and unforbidden page, together with a little poem from is pen.

CAPT. EVAN P. HOWELL,

The Napolian of the Atlanta Constitution, a man of big brain, and has wielded a powerful influence in the affairs of Georgia.

Forty years ago he was my telegraph messenger boy in the first telegraph office started in the city, and he was a bright one, and a good boy. (He may have done some devilment since then.) I remember once watching him count his days earnings, which had been extraordinarily large, two dollars and five cents, (he got five cents on the delivery of each message.) He looked up at me with eyes sparkling with inspiration, and said Mr. Sloan I am going to be a rich man. He has made the rich man, loves a good joke, enjoys life; and if perpetual youth could be preserved, think he would be willing to cast his lot with Atlanta forever.

Now my head is growing whiter,
  My days they will soon be ended;
I dropped my pail, and spilt my milk,
  And its too late to be mended.

After another forty years have sped,
  Don't think I much shall care
How this little world is wagging,
  Nor that I didn't get my share.

Got to think next world's the big one;
  My future hopes all center there,
And now wouldn't swap the chances
  For the wealth of worlds down here.

I have made right smart of money,
  But somehow never could it keep;
The thing was so slick and eely,
  That I couldn't make it heap.

I never loved the mighty dollar
  As much as what it would buy,
And I couldn't keep from spending,
  I reckon that's the reason why.

But a word more about Atlanta,
  Our grand Lady and so fair;
The crown upon her queenly head,
  The sparkling jewels in her hair.

Such beauteous face, winsome form—
  Ain't she a daizy, daizy belle?
And what her triumphs are to be,
  Is more than weuns now can tell.

She steps lightly like a fairy,
  All her movements so full of grace;

Like her namesake, swift Atalanta,
She's the champion in every race.

Who'd have guessed this cracker girl,
    The same that wore the shabby dress;
But though she was a cracker girl,
    She had a wink to business.

Now her suitors count by thousands,
    Under each one's arm they say is worn:
Ever alert to start to tooting,
    For 'tis said, each one toots a horn.

And she dances to the music,
    She cuts a caper for every blast;
Can't count the twinkles of her feet,
    Kase she flings em out so fast.

Lightly trips our Lady Atlanta,
    So lightly trips to the mazy dance;
She's the belle of all the townies,
    And she's passed the line of chance.

See her church spires point to heaven,
    She's kind and helpful to the poor;
She has a welcome for the stranger,
    Her latch string's outside the door.

Her breezes are soft and balmy,
    And her heart is ever warm and true;
You bet her dirt is good investment,
    And will bring in the revenue.

Toot your horns for the cracker girl,
    The girl who wore the shabby dress,
For now she's a Dolly Varden,
    She's a hummer, and nothing less.

PROHIBITION VICTORY IN ATLANTA, GA.

The battle is o'er, the victory won,
Prohibition has saved, saved her son;
The blue ribbon triumphs over the red,
The goggle-eye monster, drink, is dead.

The day of Jubilee has surely come,
Prohibition has slain, buried rum;
Great rejoicing throughout the camp,
Jined with the saint is the Jug-a-wamp.

Dear Atlanta, for thee there is a boom,
Blessings for which there can't be room;
Shake hands, victors, shake hands, shake,
We've fought the battle, won the stake.

One Boniface, he sot on the fence,
Reckon, bekase he'd no better sense;
And he didn't take to narry side,
So sot on the fence, sot thar, astride.

He heard Prohibition's vengeful cry,
See'd the sparks come outen her eye;
See'd her light onto the crouching brute,
But he sot thar still, sot thar mute.

Sot thar till Betsey kilt the bar,
Watched through all the howlin war;
Saw Prohibition Betsey whip the fight,
This Betsey gal--oh! wan't she a sight?

She fit and prayed, and prayed aud fit,
And, golly, didn't she make old Anti git?

Preacher folks helt Betsey's skirts;
'Tis tho't some of em tore their shirts.

Niggers jined in, throwed Betsey chunks,
Kase 'twas fore-agreed to go in hunks;
White gals pinned ribbons on their coats,
And served them lunch to git their votes.

And thus it wer old Agaric died,
Who had so long the law defied;
A rose smells as sweet by other name,
But whiskey will be drunk all the same.

For Prohibition now sound three cheers,
Let defeated Antis close their ears;
Now we'll ride upon the upper decks,
We've got our feet upon their necks.

Come world, we open now all our doors,
Prohibition will fill all our stores;
Come world, Atlanta is now the hub,
Strike the tambourine, a rub a dub, dub.

And the poor imbecile, who had no will,
His only grits from prohibition mill;
Poor devil, who couldn't pass a bar,
He too, saved by this mighty war.

Poor old Anti, whar, oh whar now is he?
Lets wait two years, then we'll see;
He may be dead, past resurrection,
But apt to hitch on another connection.

Now your Poet feels little sort of sad,
For in this victory he no part had;
No part in victory, neither feels defeat,
Till he gits short of something good to eat.

Written the morning after the prohibition victory in Atlanta, Ga. In that election I was neutral. I could not vote for whiskey, on the other hand, I thought the jug business would make matters worse.

OUR OLD CHIEFTAIN.

(Mrs. Davis offers her husband medicine.)

Pray excuse me, my dear wife,
Medicine cannot save my life;
Pray excuse me, a gentle wave,
With his enfeebled hand, he gave.

Pray excuse me from the pain,
For to no good it can attain.
Pray excuse me, was his last word,
His last speech ever heard.

Pray excuse me, I must away;
I must go, cannot longer stay.
Pray excuse me, my dear friends,
For angel spirits now attends.

Pray excuse me, oh, my South,
Last words from his mouth.
Pray excuse me, I cannot take,
Give to widows and happy make.

Pray excuse me, I cannot receive,
Let your alms orphans relieve
Pray excuse me, nor think me proud,
Want not charity, prefer my shroud.

Pray excuse me, ye who hate,
For I have been a man of fate.
Pray excuse me, ye who spurn,
For my people my heart did burn,

Pray excuse me, my country's flag,
From my South I could not lag.
Pray excuse me, if I must sever,
Forsake my country, never, never.

Pray excuse me, if I must yield
My country's cause upon battle field.
Pray excuse me, for I must bear
Her lost cause I held so dear.

Pray excuse me from the scorn,
For to entreat I ne'er was born.
Pray excuse me, I'll bear the blame,
God is my judge, I know no shame.

Pray excuse me, time shall clear
The shafts of venom I did not fear.
Pray excuse me, perhaps 'tis best;
I did my duty, now I'll rest.

Pray excuse me, if I did wrong,
In heaven above this be my song.
Pray excuse me, I must be gone;
To heaven alone shall I atone.

A clerk in the War Department, says he would not lower the flag on the death of Jefferson Davis.

THE LITTLE PURP.

There is a little purp at Washington,
   Don't know the size of his body;
We'd bet he's got but little brain,
   Yes, we'd bet a brandy toddy.

This little thing, he turns loose,
   The little fellow seems a talker;
This little fice, with noisy mouth,
   Like all his breed, he's a barker.

Like the purp that beyed the moon,
   He tries to bark at Jefferson Davis.
These fice purps will make a fuss;
   From fice purps, Lord, save us

Borrowed wit from Old Beast Butler,
   He would bury the Mexican leg;
Then would hang the Davis body—
   Little purp, he'd suck an egg.

This little would-be son of Mars—
   Underling, they call him partridge—
Without asking, tells what he would do,
   And he's never smelt a cartridge.

Dry up, dry up, you little purp,
   Your bark, it sounds too ficey;
Better wait till you are asked,
   You need both wit and policy.

THE MESSENGER OF PEACE.

A meteor flashed athwart the sky,
  A star of wondrous brilliancy;
And thousands gazed as it went,
  And thousands grieved when spent.

O'er a continent in its flight,
  South to north flashed its light,
Then sank at old Plymoth Rock,
  First landing of our parent stock.

Like Bethlehem's star, mission peace,
  The bonds of hate sought to release;
Messenger from heaven, oh, brothers, heed,
  To heal the wounds that still do bleed.

Glorious Grady, thy work was short,
  Brave and alone, with no cohort;
Like the master, sowed the seed,
  Like the master, martyr to the deed.

WHO IS POOR?

We may be poor in worldly gain,
　May be poor in what's called lucre,
Our purse strings ever were too short—
　No, we never was much on euchre.
Yet we are not without possessions
　Too rich for sordid gold to buy.
My wife is worth a nameless price,
　And we rate her none too high.
Have our boy, and he's all right,
　He's the gentleman, every inch;
He's as true as steel, and staunch,
　And we can trust him in a pinch.
Our old fiddle still is left to us,
　Have had it nigh on to fifty years;
Has been in all our ups and downs,
　And still shares our joys and fears.
Wife's old piano still sounds sweet,
　Though 'twas bought before the war;
Like Mary's lamb, has stuck to us,
　Though it has gotten many a scar.
And there's our old buckhorn pipe,
　Our consolation in every wail;
Dear reminder of former days,
　Will go with us down to the vale.
Our old horn hangs by the wall,
　And it is not unknown to fame;
Though sighing now among the willows,
　It grieves but knows no shame.

Still have our manhood, all our pluck,
    While we live we'll strike away;
If our coffers have failed to fill,
    We trust we struck for higher pay.
Then who is poor? it can't be us,
    It might be the rich to-morrow;
To leave behind what we have named,
    Can only cause us sorrow.
From the old fiddle, horn and pipe,
    From these we'll have to sever;
Wife and boy we'll meet again,
    No more to part, no more forever.
We doubt not God will care for us
    As long as he lets us stay.
So will wait and laugh, be content,
    Whilst fleeting time is called to-day.
More than all, the blessed hope
    Of Christ, through riches of his grace;
Oh! what joys we expect in Him,
    When we reach the heavenly place.

[From the Atlanta Journal.]

HOTEL POETRY.

The National Hotel.

Our friend, Major D. U. Sloan, senior proprietor of the National, has much native wit about him. He occasionally drops into poetry, which bristles with points, as note the following :

>Friend and stranger, you would do well,
>To stop at the National Hotel,
>In Atlanta, you'll see it stand
>At Peachtree crossing, close at hand.
>Stands in the center of the town—
>The business center—sets you down.
>Our doors are open night and day,
>With a welcome in the good old way.
>
>Not first-class, in a high-faluten sense;
>First-class middle-class, all intents.
>Nabob or Dude might histe his nose,
>The Peacock tribe expect to lose.
>We seek no style, we make no show;
>For paraphernalia we do not go,
>Of solid comforts have the best,
>In these good things we do invest.
>
>'Tis not glitter that makes us rest,
>The homelike fare is oft the best;
>Good appetite needs no display,
>Better tempted the good old way.

Our meals are square, the cooking good,
We put up things just as we should;
Our rooms are nice, the linen clean,
Servants attentive as ever seen.

A first.class middle-class hotel,
Where best people are wont to dwell;
Great middle-class, of all the best,
Purest, truest of all the rest;
The strong bulwarks of every land,
Their country's pillars do they stand;
They are the muscle and the brain,
Pardon us if we speak too plain.

But grand hotels there ought to be,
Grand things are pleasant for to see;
Let those who have to throw away,
Suit their fancy, for which they pay.
Must be hotels of every sort,
Hotels for purses that are short;
Ours is run on the solid plan,
The right place for the middle man.

Now, friend and stranger, what you need
Is, that your purses should not bleed.
Good fare at reasonable prices,
Not to pay for extra devices;
Then, traveler, stop with Dave U. Sloan,
Make you feel like his house is yourn;
He'll feed you and sleep you so well,
You can't keep away from his hotel.

SEWING MACHINE POETRY.

The world moves on, moves on apace,
Onward moves the mighty, mighty race;
Could our fathers, back from the grave,
See how we doth labor, labor save;
What invention for us hath won,
What half-century for us hath done;
See the great steamers plow the sea.
Long railway belting hill and lea,
Telegraph, telephone and wonder sights,
Oil, gas, and grand electric lights,
Never dreamed of fifty years ago.
But onward, flying onward, here we go,
Fulton and Morse first led the van,
This day Edison is a mighty man;
Already the world is set ablaze,
Invention seems to be the craze.
Now here comes Brosius' new machine,
For woman the best of all I ween,
To woman, 'tis the greatest blessing,
She now can sew without distressing.
No treadle now for her to beat,
No labor now for weary feet;
But touch a spring, the needle goes,
She guides the cloth, 'tis all she does;
No aching back no tired feet,
Just sitting upright in cozy seat,
A pleasant past-time 'tis now to sew,
With Woman's thanks to brother Brosius go.
All mothers will bless the Brosius name,
Their daughters will ever do the same.
Benefactor Brosius, to female race,
Thou hast solved a serious case;
God has inspired thee with this fact,
Guided thee in this genious act.

D. C. SLOAN.

ATLANTA, GA., Sept. 5, 1888.

A WORD FROM A CENSUS TAKER.

[From the Rome (Ga.) Tribune.]

Avast, There! Atlanta.

It was only by the most strenuous exhibition of nerve and obstinacy that one of Rome's most distinguished citizens escaped a forcible enumeration as a citizen of Atlanta.

The Roman (and his name would add immense pungency to this narrative) was sitting in his room at the Markham, engaged in business conversation with a citizen of Texas. There was a knock at the door, and in answer to invitation, a census enumerator walked in, took a seat, unrolled his papers and presented a blank to each of the gentlemen to be filled.

"But I am not a citizen of Atlanta," said the astonished Roman.

"That s all right," said the enumerator softly. "You are here now."

"Well, but I live in another city," said our distinguished fellow citizen.

"Yes, but you may not get back in time to get counted there," urged the Atlanta census taker.

"Oh, I live in Rome. I left there this morning. I am going back this afternoon," protested the Roman.

"Still the train might run off with you. Better let me count you anyhow."

And the stirring official, with the noble spirit of Atlanta

pulsing in his wrist, was just about to capture and appropriate to the State capital one of the best names of the mountain metropolis, when the owner of it rose up and invited the obdurate census man to go out and get a breath of fresh air.

Our citizens of Rome generally—and this one in particular—are good enough generally to divide in half and make two good citizens to the growth of other towns, and there is no place we had rather share with than Atlanta, but somehow the pride of locality is so great with Romans that none of our people care to be counted away from home this summer.

We regret now that this disposition was not published earlier for the benefit of Atlanta enumerators.

EDITORS CONSTITUTION: I read in the Constitution this morning the article from the Rome, Ga., Tribune, a thrilling episode, of how "one of Rome's most distinguished citizens, by strenuous exhibition of nerve and obstinacy, escaped forcible enumeration by one of Atlanta's census takers."

That ferocious enumerator must have been me, as the Markham House was a part of my work. The list of names and rooms were given me by the clerk, and I called to see the parties.

This particular case is adorned with so many embellishments that I fail to remember it, and must say that it savors to me much of romance. I take the occasion to beg the gentleman's pardon, if I seated myself in his room without invitation.

I beg his pardon again if I failed to distinguish him as one of Rome's most distinguished citizens. It did not occur to my obtuse mind that he was a mogul, nor the disproportion between his size and make-up, and that of my humble self. I do not remember even that his august presence inspired me with the slightest terror.

And once more I beg pardon if I handed him and the gentleman from Texas, the schedule blanks to fill, for in every other instance, I have asked the questions, and filled out the blanks with my own hand.

If the conversation occurred as related, I do not remember it, and if true, it was in pleasantry. I like a little fun, but I do assure the distinguished gentleman from Rome, that I did not seriously entertain the thought for a moment to commit a rape upon his great name, for the emolument of Atlanta. The conquest of a more insignificant name would have been attended with less hazard, and would have answered the same purpose.

Now as to the last statement, the grand climax and finale.

"When the census taker was just about to capture and appropriate one of the best names of the mountain metropolis, then the proprietor of that great name, rose up and invited the obdurate census man to go out and get a breath of fresh air."

Now that part of the story is too thin, for I reckon there are a hundred good citizens in Atlanta who know the census man referred to, and it would be hard to make them believe this census man was so forgetful of his principles, and his hereditary courtesy as to fail to return the compliment by asking the distinguished gentleman from Rome to come out, too, and breath the fresh air with him.

When he speaks of the stirring official with the noble spirit of Atlanta pulsing in his wrist, if he will allow the substitution of the word heart for wrist, the phrase will be accepted; for the census taker loves Atlanta, for forty years his heart has been with her. He knew her when she was weak, and shabby, and poor, and now adores her in her strength, splendor and greatness. He has watched her eclipse all other cities in Georgia, and now that she has promise to become the great city of the south, she has no time to rebuke the barking at her heels; but she has enough humble citizens, like her census takers, to take care of her trailing skirts.

Respectfully,   D. U. Sloan,
The Census Taker.

HON. LOGAN E. BLECKLEY,

Chief Justice for the State of Georgia, a self-made man, has ever been a hard worker, a man who stands square in his boots, and asks no favors. A man distinguished for his conscienciousness, ever willing to take the fare he offers to others; a man who knows how to imitate nothing or nobody—if not an original character, then nothing. Generous and just in all his dealings, and adorned by nature with a large share of the Christian virtues, but a man that I have thought has given more of his attention to the teachings of Moses than he has to the teachings of Christ, the greater of the two. He has been my friend for forty years, and from whom I have received many acts of kindness. A man to be admired as much for his simplicity of manner, as for the largeness of his intellect.

JUDGE BLECKLEY'S PHANTOM LADY.

O, Lady, Lady, Lady:
  Since I see you everywhere,
I know you are a phantom—
  A woman of the air.
I know you are ideal,
  But yet you seem to me
As manifestly real
  As any thing can be.
O, soul enchanting shadow,
  In the day and in the night,
As I gaze upon your beauty
  I tremble with delight.

If men would hear me whisper
  How beautiful you seem,
They should slumber while they listen
  And dream it in a dream;
For nothing so exquisite
  Can the waking senses reach—
Too fair, soft and tender
  For the nicest arts of speech.

In a pensive, dreamy silence
  I am very often found,
As if listening to a rainbow
  Or looking at a sound.
'Tis then I see your beauty
  Reflected through my tears,
And I feel that I have loved you
  A thousand thousand years.

SENATOR JOSEPH E. BROWN,

Born in old Pendleton District, S. C.; married Miss Elizabeth Gresham, who was born at old Pendleton; made the law his profession. Elected to the Georgia Senate, next Governor of Georgia for four consecutive terms. From that position he went to a Federal prison, next Chief Justice of Georgia. Elected President of the Western and Atlantic Railroad, and to the United States Senate. Senator Brown gave fifty thousand dollars to the education of the poor young men of Georgia, and fifty thousand dollars to the Baptist Theological Seminary of the South; and has, in his day, wielded more power in Georgia than any other man alive or dead.

THE POOR BOY.

Though a secessionist and a confederate soldier, I rejoice that the Union of the States has been preserved, and pray that this Union may never be severed.

If wrongs shall occur, as they undoubtedly will, from time to time, I have confidence in the people. If through party spirit, excesses and outrages are perpetrated by one section upon another, I believe a right thinking people will correct the wrongs at the ballot box.

Here is a brief history of the lives of two American Southern boys, two cousins, both with brain and brawn, the one from the hill-sides of poverty, the other from the lap of wealth; the fortune of one, that he started poor, the misfortune of the other, that he started rich.

The birthplace of the poor boy was among the backwood hills of old Pendleton District, South Carolina, near the Georgia line, and opposite the counties of Rabun and Habersham.

In his youth his parents moved over into Union county, Georgia, to a section still farther remote from the advance of civilization; here our poor boy was compeled to labor daily on the little farm to aid his father in the support of the family, and in such spare times as he could command for himself, he cultivated patches and corners of the fences for his own private means. After a time, he had saved enough of his hard earnings to purchase a pair of small steers, which he broke to the yoke.

From that small start, from that insignificant possession, sprang in his mind a great conception ; he thirsted for knowledge. He had learned from surrounding nature that from the little acorn the great oak had grown, and right upon those steers was laid the foundation for future greatness. He had determined to sacrifice the steers, earned by the many days' sweat of his face and tiresome toil of his body, upon the altar of wisdom, and resolutely and literally "steered" his course in that direction.

Watch him as he starts from the humble home of his parents, dressed in a plain homespun suit, wool hat, and home tanned shoes ; he walks behind, and with a plow line drives before him his steers. I am familiar with the roads he traveled then and in my imagination can see him now as he wends his way over the mountains, up and down the long, rough, steep, hills over creeks and rivers, on and on for more than a hundred miles. As he wends his weary, lonesome way, the passing equestrian but little dreams of the undeveloped power hidden in his humble mien ; and as he ghees, or haws his steers to either side of the road, to allow the splendid equipages to pass with their stylish occupants, unnoticed by them, could it have entered their thoughts that one day that shabby youth would be able to buy out their aggregate possessions and still have abundance left. But on and on he trudges with weary feet, intent upon one great object, to seek the temple of learning, sometimes overtaken by the darkness of night, alone, friendless and unknown, except by his steers ; at last his destination is reached he has arrived at the Calhoun academy. He soon trades his steers for eight months' board, and arranges his spare hours to labor for his tuition.

Next we see him as he sits under the shades of the great

oaks near the academy pouring over his lessons. We now introduce the wealthy cousin, who has also come to the Calhoun academy to be educated. He comes in a carriage, is dressed in broad-cloth, and has money in his pockets to spend as he likes, but withall a clever kind hearted, rollicking, friendly and talented fellow, ready for fun or for a fight at the drop of a hat, but with no incentive to self exertion, or self daniel.

Time passes and the poor boy's means are about exhausted, he must soon abandon his studies and return to his humble and obscure home, to his old time daily toils.

One day the rich cousin approaches where he sits under the trees at his books, advises him to give up the foolish idea of an education, to abandon an ambition so preposterous. He said to get an education required money, that he had already fooled away his steers; to go back home and when he got hold of another pair of steers to hold on to them; that it was not his lot in life to have an education; to be content to remain in sight of his daddy's cow pen—that he could be happy there. Said his own father was rich; that the negroes were like black birds in his father's fields; that he would have money to back him; he would go through the South Carolina college; that his career would be onward, upward, excelsior, by-G–d, and concluded by saying to the poor cousin, when I am thundering in the halls of congress where the h–ll will you be.

This was discouragement, but it did not discourage him. The poor boy returned alone to his humble and obscure home; but he had got a taste, he had learned to read, to write and cypher, and plodded on as best he could for advancement, and the rich cousin went to college, and here we draw the curtain for a season.

Nearly twenty years have passed, the scene now opens at

Milledgeville, Ga. There is an assembly of guests at the Mansion. Two gentlemen from S. C., and the once rich cousin, who is now a member of the Legislature, are the guests, and are entertained by the once poor boy, now Governor Joseph E. Brown, and his wife. They are talking of bye-gone days, and the Governor relates the story of the school boy days, and the advice of the rich cousin, who meditatively replies, "well, Joe, the changes and phases of human nature are d—n strange, arn't they? Once more we let the curtain fall.

Another season of twenty years have intervened, and the scene has shifted again. Joseph E. Brown is now a U. S. Senator, and is really a thunderer in the halls of congress, but the once prosperous cousin, where, oh where is he? The rich man's son a wanderer in a strange land among strangers, the poor boy a man of untold wealth, and upon whom all the honors of his adopted State has been heaped.

The cousin was a warm friend of mine, and a school mate, and by nature a real noble fellow; his great misfortune was that he was born with a silver spoon in his mouth. The once poor boy lives to-day in Atlanta, Ga. A phenomenal success in every thing he has undertaken, known to the world and to fame, and the best illustration of the possibilities of a poor young man, perhaps, that there is to-day in America; and no doubt that many of us, of ante-bellum times, would have been more useful citizens, and better off in the world, if it had not been for the difficulties of the silver spoon.

THE OLD NORTH STATE.

SURREY COUNTY, N. C., April 26.—Editors Constitution: The biggest man in the world was fattened in North Carolina, having reached the enormous avordupois of one thousand pounds. The anecdotest man in the world lives in that state. The largest lump of gold was found there, too, and the Buncombe part of that state is without a rival for its cabbage heads.

If you stick a pin through the map, near the coast at the northeast corner of North Carolina, about Currituck, and whirl the old North State around with your finger, you will find the southwest corner, in the circumference made will brush both the states of Maine and Florida, and that it embraces within its territory a greater variety of products than any state in North America—from cotton to tobacco, rice to buckwheat, tar to balsam, goobers to chestnuts, coal to iron, nickle to gold, corundum to diamonds—and let it be ever remembered by all Americans that it was from this old North State the first bugle notes of independence were sounded, whose clarion blast awakened this great continent, and its sound went across the great waters to the shaking up of kingdoms. Great is the old north state.

Recently I listened to a song of the old North State, rendered by a Surrey county young lady with so much patriotic pathos that I caught the inspiration. As each verse proceeded, so increased my reverence for the old North State. When Fulton county, Ga., was but a howling wilderness, when her hills were only known to the red man and her forests the habitations of wild beasts, when the great city of Atlanta was as yet unconceived in the realms of thought, Surrey county, N. C., was already settled by an intelligent and hardy race of Caucasians from the isles of Great Britain, whose decendants still hold the fort. Some of them have grown rich in lands, tenements and herediments, but still adhere to their original simplicity of manner and dress. With these unpretending people the outside fix-up of a man is not an index to his financial condition. I imagine if one of Sam Jones's spider-legged, toothpick-toed dudes were to alight about the Pilot mountains, he would be taken for some kind of a stray sea bird driven in by the storms, captured and caged by a sewing machine agent, and carried around for a nickel show.

Mount Airy, the principle town of Surrey, is located on a high ridge, in the fork made by the Ararat river and Stewart's creek, half surrounded by the Blue Ridge mountains, presenting a most enchanting view. It is already quite a flourishing town, without railroad facilities; has 1,500 inhabitants, mostly white; is remarkably well built. I noticed many handsome, even stylish residences; beautiful shades and flower gardens; large brick stores, warehouses, manufactories and tanneries; has a capital newspaper, a brassband, hotels, schools and churches. I heard modern music floating out from parlor windows; saw well dressed ladies on the streets, some even with bustles, but not of the huge proportions we often meet on the side walks of Atlanta. I met a young lady resident of Mount Airy, who had triumphantly scooped up three first honors from the different colleges, and whose artistic touch on the piano would command the admiration of Peachtree circles in the gate city of the South. Another young lady is a finished cabinet workman, as well as an accomplished musician, who handles the saw and chisel, and the piano keys with equal talent and facility, possessing a superb and cultivated voice, and is the organist of the Baptist church.

The finest wool blankets, cassimeres and jerseys are manufactured here by the Moore Brothers.

In sight of Mt. Airy is the birth-place of Daniel Boone, of Kentucky fame. His name is still to be seen chiseled out on a rock by his own hands, in the yard of the old homestead. Spending a night at the dilapidated old town of Rockford, we stabled our horse in the room of the old court house where Andrew Jackson was admitted to practice law and where he pleaded his first case. Just across the line, in Patrick county, I was pointed out the birthplace of J. E. B. Stewart, of confederate fame. In Wilkes, an adjoining county, our Governor John B. Gordon's father was born and lived for many years, and where his relative, General J. B. Gordon lived and won great distinction. From Wilkes county a part of my own ancestry came, on the Hackett side. Mt. Airy is built on a beautiful white speckeled granite rock, the disentegration of which has imparted a whitish color to the soil for miles around. This granite works up well, and there now lies at the quarry a slab, without a break, two feet wide and ninety-two feet in length.

My old friend Charley Lewis carried me out to see the celebrated White Sulphar Springs, four miles from town—a most lovely place. The hotel sits in a cove under the foot-hills of the Blue Ridge, with a lawn, covered with shade trees of ten or fifteen acres, stretching to the Arrarat river in the front. The analysis of this spring is the same as the Greenbrier, of Virginia. You can see the sulphur crystalized and encrusted upon the walls of the rock enclosing the spring. For dyspepsia, catarrh, cutaneous, liver and kidney affections, wonderful cures have been effected. Near this spring are the Blue Ridge pinnacles, said by an extensive tourist to be the greatest curiosity in North America, excepting Niagara Falls alone.

Twelve miles south of Mt. Airy is the Pilot mountain. Ascending from the lower hills that surround its base, it rises near 2,000 feet in a cone shape. When near the apex, it seems to have been cleft in twain horizontally and the segment patted up, corn-dodger fashion, into a ball, and then set back again, on top. I walked round in a well-worn path under the rim of this dome, a circumference of one mile, affording a beautiful view of the surrounding country, looked down upon a thousand tobacco farms. To ascend the top of the dome ladders have to be used, and on the very top is a patch of forest timber, about ten acres. Half way up the mountain is a bold spring of delicious water, gushing out from under the ledges. On last Easter Sunday I met on the Pilot a large crowd of Surrians, who make it their custom to spend Easter on the mountain. I persuaded some of the young folks to sing "Nearer My God to Thee," as we sat on rocks under the dome, started a club for The Weekly Constitution and promised to write an article about Surrey county for The Constitution. I cannot omit to mention an old time wooden clock, ten feet high, that I saw in a corner of a Surrey dwelling—an inimitable grandfather's clock. To how many generations it has ticked the destruction of time, and to how many more it will mark the passing hour, who can tell? Surrey county, N.

C., may have lain in a sort of comatose, or Rip Van Winkle state, in the past tense; but she has a destiny for the future, abounding in resources of wealth, the best hill lands for tobacco, the finest bottoms for cereals and grasses, with vast water powers for machinery, a great variety of the most valuable timbers, rich in mineral ores, an upright, energetic people. The Cape Fear & Yadkin Valley railroad is already in her borders, and will reach Mt. Airy in the early part of the coming year, and in the spring time of 1888 Madame Surrey will don a new robe upon her comely form, and with the horn of prosperity in her outstretched hand and maternal pride beaming in her face, will step out upon the stage and will present to the world her fair and blooming daughter, Miss Airy, whose fresh and genuine attractions will excite general applause from all beholders, and whose real charms will draw a cloud of devotees around her delightful circle. D. U. SLOAN.

HON. JOHN B. BENSON,

Or old " B," as he is familiarly known.
Born a merchant, in old Pendleton, South Carolina.
Died in the harness at Hartwell, Georgia,

His epitaph, or is to be when he shall shuffle off this mortal coil; and though there is only about one hundred pounds wrapped about him, has a soul big enough for a coil of treble the weight. My old school mate, and who accidentally killed me with a shinny stick, and for which I afterward freely forgave him, when I learned he was worse troubled about the matter than I was. There are few cleverer men in this world than old " B."

THE JUNIUS LETTERS.

In my article on old Pendleton, S. C., I referred vaguely to the old-time famous Junius' letters, the authorship of which has, for more than an hundred years, been shrouded in mystery, a mystery of the 18th century, and only paralleled in that great metropolis of the world by the mysterious murders of Jack the Ripper, of the 19th century.

My old friend, John B. Benson, has just opportunely sent me a clipping from the Hartwell (Ga.) Sun, of matter furnished by himself through its columns, that throws much light on the subject of the Junius letters. Old "B." says about the very beginning of the present century there came a man, a refugee from England, to old Pendleton, who brought with him a lot of type and printing material that had been used in London in publishing the celebrated Junius letters, and this man, John C. Miller, had been driven out of England on account of his connection with the printing of these letters.

Miller started the first newspaper at old Pendleton, and called it "Miller's Weekly Messenger—a paper 12 by 14 inches in size; and one day the old man had gone to dinner and left the forms all ready to be struck, when Tolliver Lewis, a young lawyer, stepped into the office, took out an E from the heading, and put in an A, making it read, "Miller's Weakly Messenger," and the old fellow did not find out the trick until the whole issue had been printed.

The name of the paper was some time afterward changed to the "Pendleton Messenger," and its size enlarged to 14 by 16 inches, price per annum $3.00, cash, or $3.50, credit. The press used was one that General Greene had in the Revolutionary war, and looked like an old wooden loom, such as the women used in those days, and two buckskin balls were used to ink the type.

After Miller's death, Dr. F. W. Symmes became editor of the Pendleton Messenger, and 25 years later his son, Seb Symmes, removed the old outfit to Hartwell, Ga., and together with a printer named Hagan, started the Hartwell Messenger. So the same old English type that printed the Junius letters also printed the Pendleton and the Hartwell Messengers.

Old "B." still has a copy of the original Junius letters in book form, printed in England, but unfortunately the date is torn from the front of the book with the cover. He has also two copies of the old Pendleton Messenger as far back as 1818, in good state of preservation, and the type of these old papers and the English book are the very same.

Old "B." says that the Rev. T. T. Christian, now of Atlanta, and editor of the Wesleyan Christian Advocate, learned his trade in the office of the old Pendleton Messenger; says he has seen him in the office with ink on his face, and as full of mischief as a pet coon. "I could tell a good one, too, on Tom, about a speech he tried to make at the old Pendleton Academy, but won't now," says Bob Thomson, editor of the Keowee Courier, who has made his mark in South Carolina. Tom H. Russell, who was the best speller in town and who could read anybody's writing, also learned there, and says he used to take orders to the store from John C. Calhoun over to Tom, to have them deciphered, so he could fill out the orders. Mr. Calhoun wrote an awful hand.

The following short sketch of the Junius letters are so interesting that I give it to my readers:

"Junius" was the signature of an English political writer, the author of the letters which appeared in the "London Public Advertiser," between January 21, 1769, and January 21, 1772. Henry Woodfall was the publisher of the Public Advertiser, and every means were used to induce him to divulge who Junius was, but without success.

These letters, directed against the ministry and the leading public characters connected with it, contain some of the most effective specimens of invective to be found in literature. Their condensed and lucid diction, studied and epigrammatic scarcasm, dazzling metaphors, and fierce and haughty personal attacks, arrested the attention of the government and the public. Not less startling was the immediate and minute knowledge which they evinced of court secrets, making it believed that the writer moved in the circle of the court, and was intimately acquainted, not only with ministerial measures and intrigues, but with every domestic incident. They exhibited indications of rank and fortune as well as scholarship, the writer affirming that he was "above a common bribe" and "far above all pecuniary views." When Woodfall was prosecuted, in consequence of Junius' letter to the king, the author promised to make restitution to him of any pecuniary loss. The authorship of Junius was the greatest secret of the age. Every effort that the government could devise or private indignation prompt was in vain made to discover it. The Earl of Mansfield and other legal advisers of the crown had many consultations as to how this "mighty boar of the forest," as he was called by Burke, could be most adroitly ensnared in the network of the law. The host of enemies whom he aroused in every direction were eager in plotting schemes for his detection. But, aware that his power and perhaps his personal safety depended upon concealment, he continued to astonish every one by his secret intelligence, and to assail the government with undiminished intrepidity and rancor, revealing his apprehensions and precautions only in his private notes to Woodfall. His security was doubtless due in large measure to the forbearance and honor of this publisher, who followed strictly the imperative and precise orders of his correspondent.

Sir W. Draper, who entered into controversy with this unknown adversary, was in the end overmastered and reduced to mere humble complaint and confession. The Duke of Bedford, Lord Mansfield, and the Duke of Grafton, all measured intellectual lances with Junius, but were made to writhe in ignominious defeat.

Who the person was who thus foiled the scrutiny of his own age has been the subject of more than one hundred volumes and pamphlets. Efforts have been made at different times to identify him with no less than forty eminent Englishmen and Irishmen, and while it may be put down as supported by the best evidence that the author was Sir Phillip Francis, still it has not yet been demonstrated beyond a doubt, and to-day the question, "Who was Junius?" remains unanswered.

DR. H. V. M. MILLER.

Born in old Pendleton District, South Carolina, grew up to manhood in Rabun County, Georgia. Graduated in medicine in South Carolina, completed his study of medicine in Europe. Settled in Cassville, Georgia, there entering politics, he became known as the Demosthenes of the mountains. Was a surgeon in the Confederate army, has been professor of medical colleges in Memphis, Tenn., in Augusta, Ga., and in Atlanta, Ga. An editor of medical journals, and United States senator. A fine speaker, a man of great gifts in conversation, and one of the best read, and best informed men in Georgia.

THE OLD STONE CHURCH.

About two miles out from old Pendleton, S. C., in the woods near a country road, stands the crumbling walls of the old Stone church, and hard by the entangled vines, the old cedars, and other decaying evergreens, grim sentinels in the dilapidated old graveyard, the whole presenting a wierd and desolate scene.

This old Stone church was built by General Pickens directly after the Revolutionary war, as a Presbyterian church. About 1845 the walls fell in, and the old church has long since been abandoned.

Many of Pendleton's first citizens are buried there. The father and mother of my old friend, John B. Benson, lie there. The remains of Colonel Bynum, who was killed in a duel by Colonel Ben F. Perry, repose there.

I have a vague remembrance of many strange and thrilling histories, and legends connected with this old Stone church, but am not sufficiently posted as to the facts to attempt to relate them here.

I would have liked to have gathered many of its histories and present them here, but have failed to do so. Indeed, if I ever attempt to write another book I would be delighted to make the whole subject upon old Pendleton, its great district.

and its people, who have lived and died and who have gone out from her borders, making their impress upon other sections of the South, for they are to be found in every part of our sunny land.

Right here I am reminded of some of old Pendleton's people, well known in this section, who have not been mentioned: Mr. I. O. McDaniel, the father of Governor McDaniel, Judge Hutchins, the father of the present Judge Hutchins, of Gwinnett county, and the Hon. W. T. Smith, of Gwinett, Mr. Ed. Werner, of the Georgia road, the printer Bridwell—all came from old Pendleton. So did the great Dr. Lewis, the father of the State road, and to tell of all, would require a book for that purpose alone.

JUDGE WILLIAM LOWNDES CALHOUN,

President of the Board of Trustees for the Confederate Soldiers' Home, Lieutenant Colonel Georgia Volunteers, President of the Confederate Veterans' Association of Atlanta Ga., Ex-member Legislature, Ex-Mayor of Atlanta, and the present Ordinary of Fulton County. A man of extra fine executive ability, and one whom the people delight to honor. And whilst Henry W. Grady may be called the projector of the Confederate Soldiers' Home, President Calhoun has been the perfector of the work; it has been to him a work of love, and to which he has devoted a great deal of his valuable time, without charge. He has superintended and directed every item of the work on the Home, and if it is ever used for that purpose, there ought to be two busts placed in the niches, one on either side of the entrance to the Home, one of H. W. Grady, the other of W. L. Calhoun.

THE CONFEDERATE SOLDIERS' HOME.

A year or two before his death, Henry W. Grady and other patriotic citizens of Atlanta, conceived the idea of a home for the old and helpless veteran soldiers of Georgia.

Mr. Grady entered into this noble work with all the ardor of his enthusiastic soul. Others soon caught the spirit, and warmed up to the aid of this most commendable purpose. The first thousand dollars was contributed to the Soldiers' Home from a gentleman in New York. This was followed by several subscriptions of a thousand dollars each from wealthy gentlemen in Atlanta. Then many citizens of Atlanta, and other parts of the State, subscribed smaller amounts, to build the Home for the old soldiers.

When a considerable sum had been raised, a board of trustees, consisting of thirty of the best men, and from different parts of the State, were elected, who purchased 120 acres of land in sight of the city of Atlanta, selecting a beautiful site on an eminence surrounded by a grove of majestic forest trees. An architect was employed, a suitable plan designed, and a contract for the building was let. The result is that an imposing wooden structure has been erected, containing 67 rooms, spacious halls and delightful verandas, making a grand and convenient Home for the old soldiers.

A street car line has been extended to the very doors of the Home. Drives have been graded through the grounds, and orchards have been planted. A force pump now throws the water from a clear spring up into lofty towers, which is conducted thence into convenient parts of the building. Laundry and bath rooms have been arranged, and the most convenient pantry and safes. A splendid range and boiler and complete outfit of cooking apparatus stands in the roomy kitchen ready for use (a present); a heating arrangement has also been put in the building (a present); both presents from parties in other States; a nice organ (a present from an Atlanta firm), and parties from another State offered to put in a gas plant worth $2,000. This last gift, perhaps, is lost, by the delay of the last legislature to accept the property, besides a crop on the land and a year to the old soldiers.

This valuable property is all paid for, nor has a single incumbrance. During the last legislature the whole outfit was tendered by the trustees, as a gift to the State of Georgia, with the single condition that the State accept and agree to take care of her old and helplesss soldiers for a period of 25 years, and after the expiration of that time, the property should belong to the State to dispose of as she thought proper.

A bill was introduced in the house to that effect, and was referred to the finance committee, where it seems to have nodded a few times, and finally, just before adjournment of the body, to have dropped off on the table into a dead sleep, and if, like Rip Van Winkle, it shall ever awake again, can but rub its eyes and discover that much valuable time has been lost.

This property, so nobly offered as a gift to the State of Georgia, is to-day worth one hundred thousand dollars, and the best real estate men say at the expiration of the 25 years it will be worth from three hundred thousand to half a million dollars.

It is also estimated that an average of ten thousand dollars per annum for the 25 years, or two hundred and fifty thousand dollars in the aggregate, will be sufficient to conduct the Home through the period named. Of course, the largest amount would be required the first half of the time, as in the last half their members would greatly diminish, and in the last years there would be few, if a single one, left.

Thus it will be seen, the proposition is a clear and incontrovertible one: that the State, in accepting this valuable gift from the donors is presented the opportunity to care for her old soldiers by the mere loan of the money, with the absolute certainty of having the entire principal reimbursed, and the probability is that not only will the interest be returned in the end, but a handsome profit on the investment.

Every State, North and South, with but one or two exceptions, have their homes for their old soldiers, and have secured them by an outlay of money, and still the great State of Georgia hesitates when she is offered the singular opportunity to provide for her old veterans without cost.

The adjourned session of the legislature meets within a few days. What will they do with the home? This is the question. Much valuable time has already been lost. If they refuse to accept the noble gift, there is but one legitimate course left to the trustees—to sell the property and return the proceeds to the contributors.

Can the bill longer sleep in the committee room? Will it not be awakened from its long sleep on that committee table? Will it not be sent back to the house for a hearing? Will the people never know who are its friends and its foes? If there are reasons why this seemingly noble work should die, let the people hear the reasons. If there is argument why the Home should not be received by the State, let it be ventilated. Let the people hear. Let the silence be broken.

CONCLUSION.

In winding up this my first and, in all probability, last attempt in the manufacture of literature, I am free to confess the many imperfections in the little book (for indeed I have discovered not a few of them myself); and no doubt some of my conceptions may be objectionable to some of its readers I answer, the only trouble on my part was want of better sense, and if perchance I had got the whole thing plumb right, it would not have suited everybody. Even honest people may differ, see things in different lights and shades from different standpoints, and though we may disagree in some things, we can still be friends; and for the sake of peace, I will go so far as to say that you may be right, and I wrong. I only claim my convictions, and accord the same to you. To a great degree we are all creatures largely influenced by generations, surroundings and circumstances; our teachings have much to do with our likes and dislikes, with our prejudices, for or against.

I have concluded in my declining years, that whilst I look upon my fellow man as a very wonderful being, and am constantly amazed at his clevernes, startled at his cunning ways, his marvelous inventions, and the vastness of his worldly wisdom, yet I have discerned that there is a limit to his capacities, and to his accomplishments, as there is to his temporal life, and that after all his seemingly big ways and doings, he is, at least, but a very simple and foolish creature about some of the most important things, and that some of the very wisest of the world, are to-day engaged in the silly and unprofitable employment of trying to stop up the little leaks of life, and leaving open the great bung-holes of eternity; and in summing up the whole, there is very little difference between the worlds wise man, and its fool, and that the history of both may be summed up into blunders, one half misdeeds, and the other half mistakes; and I have even thought it possible that more of the world's fools may be saved in the end than its wise.

The successful man of the world is, by common consent, considered the wise man, and upon him are the honors and the adorations of the world heaped. Although the very program, in a special sense, is in direct defiance of the written laws of God.

On the other hand, the same law condemns the sluggard, but the true wisdom is clearly given—given too plainly for mistakes—and is contained in the little text, "Seek first the kingdom of God."

How plain; who can mistake its meaning; and shall not all men be held responsible for its infringement. It clearly applies to all—the wise and the foolish, the rich and the poor, high or low. We are all amendable to this commandment; and why cannot all sane men recognize the unalterable fact, that the greatest of all wisdom, is to seek the greatest amount of good, and that good that will endure for the greatest period of time; which can alone be found within the pale of the permanent plan of salvation, as promised in the written word of the allwise Creator, the maker of the heavens and the earth.

Then as we are all in a common trouble, and the difference in our temporal condition is of such small moment, why should not we all seek to help one another; as all shall need mercy, why not be merciful; as all shall need friends, why not be friendly; why should the humble hate the proud and self con-

conceited, for all too soon they will have their day of reckoning; why should the rich and strong despise the poor and weak, for their own day of helpless poverty is but postponed.

What man with common sense, who will stop a moment and think, can fail to conclude, from his own earthly observation, that old Solomon was right, when he pronounced the fleeting things of this earth all vanity. We must all leave the world and its folly far behind us.

A certain hard student in his youth, and an able jurist in his maturity, is accredited with the saying, "That the next best thing to religion is fun;" and he was perhaps not far wrong, though like the poet who wrote that incomparable song, "Home, Sweet Home," and was said never to have had a home, we have thought this jurist and student must have had large imagination, for in his studious youth he had but little time for fun, and under the arduous duties of the ermine, less time for religion.

I have thought that the man who loves his God and his fellow man, cannot be adverse to fun, harmless fun. Tying a tin can to a dogs tail in wanton fun; to fight dogs and chickens, is cruel fun; to profane the Sabbath with unrighteous merriment, is sacreligious fun; but to surprise suffering humanity with acts of kindness, and with timely aid, is heavenly fun. The frolics of the lamb and kitten, are innocent fun; the birds, when they flit so merrily among the green boughs and chirp and sing, are having their fun; nature itself clappeth her hands for joy, and this is the kind of fun we mean. The Prodigal's brother hated fun.

The man who hopes for heaven ought to be merry, and the merry man maketh his neighbor merry. A good, genuine, hearty laugh is the sign of a happy man. But there is a wanton, wicked chuckle, in which there is no fun, that scorches like fire and nips like hoar frost, a chuckle that sarcasts its hisses from lips of venom.

I rather think there will be fun in heaven. There is a pleasing sensation of merriment to me in the idea of being freed from all temptation to evil and sin; the very thought sends up a fountain of joy swelling from the heart. I have heard the extatic laugh of the happy Christian, as the soul soared away from the sin-striken world, into the purer atmosphere of the holy heavens.

There may be no fun in heaven, but I feel sure that there is none in hell—none of the kind that I think I would like. We get a little foretaste of sweet, innocent fun on this earth, and I think there will be oceans of it above—rejoicing, praising, laughing—lots of fun, eternal, righteous fun.

I think there is no evil on earth, except by the abuse or connivance of man, that all things God allows on earth hath some good purpose and benefit for man. Fire is good, but will burn; water is good, but will destroy; dynamite is good, but hath the elements of death and destruction; whiskey is good, but will ruin; the devil himself is good, to warn men to flee from the wrath to come—to be not of him or like him : sin is good, to show the contrast from righteousness, that all sensible men may be taught to make their choice between the two.

Reform is a matter of grace, accomplished through reason and conviction, and consequent upon love, teaching, prayer and waiting. It is not of force or the bayonet; to be growing and enduring, must become a principle.

Teach the people to avoid all the dangerous elements in life, to trust in God, the giver of all good, to laugh and be merry, and to love all innocent fun.

Friendly reader, I am possessed of but little of this world's wisdom, wealth, or fame, yet I have managed to keep reasonably happy, and moderately contented; have had considerable fun in my day—some wanton and wicked fun, and some innocent fun. I have repented of the first, and rejoice in the last. The innocent will be accredited to my permanent account, and the other forgiven, on account of the over and abundance of grace ever ready to be poured out upon those who ask for it.

In these unpolished pages, it will be easily discerned that the writer is not averse to fun, but do not claim by any means, that my efforts here will stand the test of innocence; yet I am consoled with the consideration that as I am of the earth, still earthy, I might have been engaged in some other worse devilment than in the writing of these pages.

Brother and sister, I am daily becoming more and more impressed that we are living in a wonderful age; I am impressed with the idea that the world is rapidly approaching its last and culminating epoch. First, the dismal, the silent age, then followed the sluggish, the fogy age; and now in this nineteenth century comes the butterfly age; and this butterfly age, I opine, will be the brief age, and then the millenium.

When the butterfly season is over, then the follies of the world will cease and the people will return to reason and to God. The flow of the two streams will be reversed, when the stream of unrighteousness shall fail from the drouth that shall fall on the mountains of evil, and a great stream of righteousness will flood all the valleys of sin. I think the day is not far distant when the people shall become convicted of their high-handed disobedience and ingratitude toward a loving God; that it will not be long till the veil that now blinds their eyes, will be lifted, and that they will with wonderful accord seek to serve the only true and living God; that the day is not far away when they will cry out mightily, "The Lord, the Lord, He is the true God."

I am impressed that the religion of Jesus Christ is preparing for an amazing forward movement: that there are mysterious movements about to be executed on earth's chess-board; that the inscrutable hand of Divinity is already quickening his dealings with a long-rebellious world. I have even thought that the Sunny South might become the favorite field for the advance of this great and glorious work, and have in my imagination pictured our own bright Atlanta as a central or distributing point for the great revolution, the great reformation—as a sort of new Jerusalem, a city set upon a hill. I have also imagined that our women were going to take a prominent part in the glorious work (I don't mean to preach), for I believe we have got more real, genuine Christian women in the South than any other portion of this green earth, and not a few good men; and I believe Atlanta has got more than the average of both good men and women.

Now learn the parable of the fig tree: "When his branch is yet tender, and putteth forth leaves, ye know that the summer is nigh." "But of that day and hour knoweth no man, no, not the angels in heaven, but my father only."

I do not wish it inferred from anything I have said, that I have spoken aught in envy against the rich of earth, for I believe an honest rich man is as good as a poor man, if he loves God. My idea of his condemnation comes from the scriptures, which says, "How hardly shall a rich man enter the kingdom of heaven."

I believe there are rich Christians, and that it is scripturally legitimate to make money, but that every man will be held accountable for his stewardship, and for the disposition which he makes of his means. What I mean to say, is, that every human being, whether in wealth or poverty, fame or obscurity, power or imbecility, must subsequently subserve to the will of a patient and omnipotent Creator; that he surely will be magnified in the sequel, and that every soul failing to recognize this incontrovertible truth, will have committed the unpardonable sin and inexcusable folly of the lost; and that when the greatest of all days shall come, there will be an host who will be appalled at their own neglect and worse than folly when the decided facts shall glare upon them that they chose the chaff in the world, and spurned the wheat.

Money should be regarded only as money, not simply for self-glory, but for God's glory, to honor him with; and in honoring God, man is but honoring himself in every true sense. Wealth should aid in advertising the great plan of a world's salvation. To worship the stuff, or its purchase, is worse to-day than the calf worship of the Israelites, and to use it to crush out God's poor is certain damnation to the oppressor.

To make money honestly, is right, but to be poor is not necessarily a crime, for the angels of heaven keep company with the righteous poor, and the legitimate heir of heaven watches over them in deepest sympathy as they pass through the valley of the shadow of death, as he once did himself, for he is acquainted with their griefs and has tasted of their sorrows.

I earnestly believe that poverty, afflictions and trials, have been to me a priceless boon, and that they have been sent in love; and I am trying to submit, and even to learn to kiss the chastening rod, and still be happy. I know that he who holds my destiny is good, wise and merciful; all nature tells me this is truth, and though his ways are past finding out, yet I shall trust him.

I know in a short time I shall be summoned to his presence, to stand my trial before his unerring tribunal, and have already sent in my plea of guilty, and have placed my case in the hands of an advocate who has never been known to fail in his courts, as far as I have ever heard of, and through messengers that I do not dare to doubt. I am promised an acquittal, and a free pardon, and not only so, but I have got word (and I believe it true) that there is an inheritance reserved and waiting for me, worth more than this whole world, and that can never be taken from me again, but will endure when this world is blotted out; and more than that, that I am to occupy a social position among the first families of the universe, and shall be allowed in the very royal presence of the King of all Kings. What more can we crave?

I shall endeavor to calmly await my summons, and the fulfillment of the promises. I would neither hasten nor stay the time. I want the Lord to direct the whole matter, because I have made so many mistakes and blunders. I envy no man. his posssessions, his temporal power, or his worldly fame, but I do feel for poor suffering and sinful humanity.

Wouldn't I make money if I could? Why, yes; I reckon I would, if I could make it honestly: and if it didn't make a fool of me, I feel like I would use much of it to alleviate the sufferings of humanity—at least, I feel so now. I think it would be to me the greatest pleasure that I can imagine, to help my poor, tottering fellow-man through the world, and on to heaven.

In my article on "Prohibition in Atlanta," it reads about the ministers, that " 'twas tho't that some of 'em tore their shirts." I want to say now, What if they did? They didn't care. If they thought they were doing their duty, they didn't care if they tore a dozen old shirts. I have no fault to find with them if they did tear their shirts; and sometimes, when I see what fools some men make of themselves about liquor, I feel like tearing up several shirts myself.

If I have said anything to offend our colored brother, I have not said it through ill will. I like him, and I claim to be his friend; but I mean just what I say. He can't rule here, and he nor no other fellow needn't try to write it down on the bulletin board that way. The thing can't be did. The best thing he can do, is to be content to sit at the second table. He can have good fare, but he's got to take the second table, exceptin' the Lord says so. Let him educate, get all the wisdom he can, make money, christianize, go spread the gospel in his own benighted country, send his young men and women there to enlighten, as they have been enlightened here, and the day came when they will have over there a country even bigger and better than the white man's America. If the chemical composition of his skin has been a little more flavored than ours, the Lord did it; but if he fills his place here, and is received above, I have no doubt he will then be as good as the best. There will be no difference there.

I have made a few cuts at our brethren across the north line—not in anger nor in hate, but in truth, as I understand it. No doubt they can point out some ugly wrinkles in us, too, which they do not hesitate to do. We all know that we have our faults, so let us forgive and let byegones be byegones; let us learn to understand and love each other better; et them come down to see us again, leaving their guns at home; come in peace, and bring their machines, their brains and their money with them. Let them come; we have the country. Our genial sun will warm up their hearts, and if they will so come, we will receive them with open arms. Some folks who read this book may not be democrats. Well, there is no reason for a fuss about that. If I was born with hair on my head and a democratic seed inside, and you were born with or without hair on yours and a republican seed inside, why let them both sprout and grow. The Lord, who giveth the increase, will select the timbers, when he needs them, from the forests, and use them as he likes.

Our readers may discover some inconsistencies in our writing. We shall not be surprised if they do, for our whole life has been made up of inconsistencies. We have endeavored to hide them as much as possible, but they would crop out in spite of us. Sometimes we feel one way, then another; sometimes we see one way, then another. We hope all such failings will be overlooked, as we were born with this weakness, and have never fully recovered from the disease.

I did think of having my book inspected by an expert, to have it dressed up so as to make a more respectable appearance; but, somehow, I could not bear the idea of showing off in borrowed plumage. I like store goods, and would like to have them tailor made, if my finances would afford it; but I prefer to patch up and wear my old duds, rather than to shine out in the robes of my more fortunate neighbor.

I might have done the job a little better myself, if I had taken more time and pains; but like the fellow who was going to be hung, I got impatient, and wanted the job over with. So, reader, if you find fault with the grammar or diction, you are at liberty to correct it to suit yourself; if the trouble should be in the spelling or the punctuation, then you may jump on the printer, for he was paid for that part of the work; but if you find any whole-cloth lies in the book, then you may put the blame on me.

To the large number of friends who have subscribed so promptly for the book, and to whom I have sold at least half of the one thousand copies before they come from the press, (and I consider this the more remarkable from the fact that the author is entirely without literary reputation,) and for the esteem and kindness of all these friends, I cannot find words adequate to express to them my feelings of gratitude. I only feel mortification at my limited capacity to afford them something more worthy of their attention, yet I feel sure that if I have failed to enlighten them, I have succeeded in furnishing them with a dollar's worth of fun.

There are other matters I would like to talk about in this, my conclusion, but if I make the book any bigger my printers will not allow me to sell it for a dollar; so, with my best wishes to every reader, I bid each one adieu, with the request that he be good to himself and to his neighbor, love the Lord, and so spin out my last word, as I shall the last moment of my unprofitable life, to an e-n-d.

D. U. SLOAN, Jr.,

Professor and Proprietor of Sloan's Atlanta School of Telegraphy, Atlanta, Ga.—a competent and thorough instructor in the art of telegraphy. His school is now in its ninth year, and is the oldest continuous School of Telegraphy in the South; and it has sent out its graduates throughout the country, who are to-day occupying positions of trust and profit. Professor Sloan is a young man of irreproachable character, diligent and conscientious in his efforts to instruct.

D. U. SLOAN, Jr., } D. U. SLOAN, Sr.,
Professor of Telegraphy. } AND { Manager.

SLOAN'S
ATLANTA
SCHOOL
OF
TELEGRAPHY.

ATLANTA, - - - - GA.

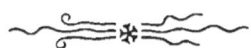

THE OLDEST TELEGRAPH
INSTITUTE NOW IN OPERATION
IN THE SOUTH.

OUR STUDENTS

Are instructed in the management of instruments, batteries and wire connections. We prepare them to send and receive both Commercial and Railroad Telegraph Business, and graduate them when they can copy from the sounders correctly twenty-five words per minute.

OUR MANAGERS

Are experienced railroad men, and familiar with the duties required of railroad employes.

Colonel D. U. Sloan, our manager, is widely and favorably known; was Atlanta's first

TELEGRAPH OPERATOR;

"an old-timer," forty years ago, and more recently for a number of years, agent and operator for the R. & D. R. R., assisted by his son, who afterwards was Professor of Telegraphy in "Moore's Business University" for four years, and since then has been Principal of Sloan's Atlanta School of Telegraphy. He has proved himself a successful teacher of others, as his numerous graduates who are now filling positions of trust and profit, will most cheerfully attest.

BUSINESS PROSPECTS

For Telegraph Operators were never brighter than now. The great number of railroads in operation, the many new ones in construction, the rapidly increasing business of the "New South" will employ a vast army of Operators, and from the present outlook, the day is far distant when an expert and reliable Telegraph Operator will fail to command good remuneration for his services. Besides, there is no trade or profession in this land that can be acquired at so little expense of money and time that pays so well. Telegraphy is a good business for the poor boy or girl, and might prove a blessing for the children of the rich to fall back upon, in case their wealth should take wings.

QUALIFICATIONS.

Any young person of either sex, with bright mind and ordinary English education, is qualified to make a successful operator.

TIME REQUIRED TO LEARN.

The average time required to learn Telegraphy in our School has been from three to four months, owing to the aptness and application of the student.

If you desire to learn Telegraphy, pay no attention to the boycotters, the disciples of the "Telegraphers' Brotherhood" or to the O. R. T'.s, who are sworn to do all in their power to keep you from learning Telegraphy, by fair or foul means. They are not your friends, and their object is obvious. Come and investigate our school and be your own judges.

OUR REFERENCES.

Prof. Moore, Moore's Business College, Atlanta, Ga.;
Prof. Sullivan, Sullivan's Business College, Atlanta, Ga.;
J. M. Stevens, Man'r. W. U. Telegraph Co., Atlanta, Ga.;
A. N. Oldfield, Electrician, Atlanta, Ga.

OUR RATES REDUCED.

To beginners, first month. .... $ 15 00
To " second month 10 00
To " third month 10 00
To " fourth month 10 00
If longer time is needed, per month.................... 5 00

OUR BUSINESS HOURS.

Nine o'clock A. M. to 12 o'clock M.; 2 o'clock P. M. to 4 o'clock P. M. Besides a NIGHT CLASS OF TWO HOURS, for those who cannot attend the Day School. Rates—$10.00 for first month, and $5.00 for each succeeding month. Board can be had in the city at from $12.50 to $15.00 per month.

TESTIMONIALS.

What has been done for the young men signed below, can be done for others. Could give many testimonials, if we had space.

BELMONT, N. C.

Professor D. U. Sloan: I take great pleasure in testifying in favor of your School of Telegraphy, where I received my instruction. I will advise all who wish to study Telegraphy to go to you. I am now agent and operator at this place, with a good salary.

WILL B. PRUETT.

ATLANTA, GA.

Professor D. U. Sloan: My sincere thanks for your good teaching. From your School I accepted a position on the East Tennessee road, at Baxley, and already I have been promoted to a better position on the same road in Atlanta, and with increased pay. I will ever hold up your School to those who wish to learn Telegraphy.

JAMES BARNWELL.

ATLANTA, GA.

Professor D. U. Sloan: To all who contemplate the study of Telegraphy, I most heartily recommend your School, where I received my training. I am now "Train Despatcher" for the R. & D. R. R. in Atlanta, and receive a salary of $100.00 per month.

B. F. MARTIN.

ADDRESS,

D. U. SLOAN. Manager
Sloan's Atlanta School of Telegraphy,
ATLANTA, - - - GA.

www.ingramcontent.com/pod-product-compliance
Lightning Source LLC
Chambersburg PA
CBHW021406230426
43666CB00006B/652